Alexander Bain

Education as a Science

Alexander Bain

Education as a Science

ISBN/EAN: 9783337035679

Printed in Europe, USA, Canada, Australia, Japan

Cover: Foto ©Paul-Georg Meister /pixelio.de

More available books at **www.hansebooks.com**

BY

ALEXANDER BAIN, LL. D.

PROFESSOR OF LOGIC IN THE UNIVERSITY OF ABERDEEN

NEW YORK

D. APPLETON AND COMPANY

1897

PREFACE.

IN the present work I have surveyed the Teaching Art, as far as possible, from a scientific point of view; which means, among other things, that the maxims of ordinary experience are tested and amended by bringing them under the best ascertained laws of the mind.

I have devoted one long chapter to an account of the Intellect and the Emotions in their bearings on education. The remainder of the work is occupied with the several topics more specially connected with the subject.

There are certain terms and phrases that play a leading part in the various discussions ; and to each of these I have endeavoured at the outset to assign a precise meaning. They are—Memory, Judgment, Imagination, proceeding from the Known to the Unknown, Analysis and Synthesis, Object Lesson, Information and Training, doing One Thing Well.

A separate consideration is also bestowed on Education Values, or an enquiry into the worth of the various subjects included in the usual routine of instruction ; the largest amount of space being given to Science.

Under the designation—Sequence of Subjects (Psy-

chological and Logical), a number of important matters
are brought forward, it is thought, in an advantageous
way.   In the first place, we are interested to know what
is the order of the unfolding of the faculties, and what
influence that order should have in the arrangement of
studies.   Such is the psychological question.   In the
next place, there is a sequence growing out of the dependence of the subjects themselves; which in most
cases is plain enough, but occasionally becomes perplexed by disguises.   This I call the logical or analytical
problem of education.

These preparatory matters being disposed of, the
main topic—the Methods of Teaching—is entered upon.
After adverting to what concerns the first elements of
Reading, I proceed to the delicate question of the commencement of Knowledge teaching.   It is here that we
are introduced to the Object Lesson, which, more than
anything else, demands a careful handling; there being
great apparent danger lest an admirable device should
settle down into a plausible but vicious formality.   The
latter part of this chapter treats of the methods applicable to Geography, History, and the Sciences.

The Mother Tongue has a place appropriated to
itself.   Everything that relates to it as an acquirement
—Vocabulary, Grammar, the Higher Composition, and
Literature—is minutely canvassed.

A chapter is assigned to an estimate of the value of
Latin and Greek at the present day.   The provisional
arrangement whereby the higher knowledge was for centuries made to flow through two dead languages should

now be considered as drawing to a close. The question then arises whether any new sphere of utility has been discovered for these languages, sufficient to reward the labour of their acquisition when their original purpose has ceased. On the assumption that the present system must sooner or later be changed, I suggest what I consider to be, in relation to the higher studies, the curriculum of the future.

On the wide subject of Moral Education, the plan adopted is to bring into prominence the points where the teaching appears most ready to go astray. As respects Religion, I have principally confined myself to the connection between it and moral instruction.

A short chapter on Art teaching endeavours to clear away some prevailing misconceptions, especially in the relationship of Art and Morality.

The general strain of the work is a war, not so much against error, as against confusion. The methods of education have already made much progress; and it were vain to look forward to some single discovery that could change our whole system. Yet I believe that improvements remain to be effected. I take every opportunity of urging, that the division of labour, in the shape of disjoining incongruous exercises, is a chief requisite in any attempt to remodel the teaching art.

ABERDEEN : *November* 18, 1878.

CONTENTS.

CHAPTER I.

SCOPE OF THE SCIENCE OF EDUCATION.

CHAPTER II.

BEARINGS OF PHYSIOLOGY.

CHAPTER IV.

TERMS EXPLAINED

CHAPTER V.

EDUCATION VALUES.

CHAPTER VI.

SEQUENCE OF SUBJECTS—PSYCHOLOGICAL.

CHAPTER VII.

SEQUENCE OF SUBJECTS—LOGICAL.

CONTENTS.

CHAPTER VIII.

METHODS.

2

GEOGRAPHY.

HISTORY.

SCIENCE.

Arithmetic.

CHAPTER IX.

THE MOTHER TONGUE.

CHAPTER X.

THE VALUE OF THE CLASSICS.

CHAPTER XI.

THE RENOVATED CURRICULUM..

CHAPTER XII.

MORAL EDUCATION.

CHAPTER XIII.

ART EDUCATION.

CHAPTER XIV.

PROPORTIONS.

APPENDIX.

FURTHER EXAMPLES OF THE OBJECT LESSON.

PASSING EXPLANATIONS OF TERMS.

EDUCATION AS A SCIENCE.

CHAPTER I.

SCOPE OF THE SCIENCE OF EDUCATION.

THE scientific treatment of any art consists partly in applying the principles furnished by the several sciences involved, as chemical laws to agriculture ; and partly in enforcing, throughout the discussion, the utmost precision and rigour in the statement, deduction and proof of the various maxims or rules that make up the art.

Both fecundity in the thoughts and clearness in the directions should attest the worth of the scientific method.

DEFINITIONS OF EDUCATION.

First, let me quote the definition embodied in the ideal of the founders of the Prussian National System. It is given shortly as 'the harmonious and equable evolution of the human powers ;' at more length, in the words of Stein, 'by a method based on the nature of the mind, every power of the soul to be unfolded, every crude principle of life stirred up and nourished, all one-sided culture avoided, and the impulses on which the strength and worth of men rest, carefully attended to.' [1] This definition, which is pointed against narrowness

[1] Donaldson's *Lectures on Education*, p. 3b.

generally, may have had special reference to the many
omissions in the schooling of the foregone times: the
leaving out of such things as bodily or muscular training ; training in the senses or observation ; training in
art or refinement. It further insinuates that hitherto
the professed teacher has failed to do much even
for the intellect, for the higher moral training, or for
the training with a view to happiness or enjoyment.

Acting on this ideal, not only would the educator
put more pressure altogether on the susceptibilities of
his pupils : he would also avoid over-doing any one
branch ; he would consider *proportion* in the things to
be taught. To be all language, all observation, all
abstract science, all fine art, all bodily expertness, all
lofty sentiment, all theology, would not be accepted as
a proper outcome of any trainer's work.

The Prussian definition, good so far, does not readily
accommodate itself to such circumstances as these :—
namely, the superior aptitude of individuals for some
things rather than for others ; the advantage to society
of pre-eminent fitness for special functions, although
gained by a one-sided development; the difficulty of
reconciling the 'whole man' with himself; the limit
to the power of the educator, which imposes the neces
sity of selection according to relative importance.

Although by no means easy, it is yet possible to
make allowance for these various considerations, under
the theory of harmonious development ; but, after the
operation is accomplished, the doubt will arise whether
much is gained by using that theory as the defining fact
of education.

In the very remarkable article on Education contri•

buted by James Mill to the 'Encyclopædia Britannica,' the end of Education is stated to be, 'to render the individual, as much as possible, an instrument of happiness, first to himself, and next to other beings.' This, however, should be given as an amended answer to the first question of the Westminster Catechism—'What is the chief end of man?' The utmost that we could expect of the educator, who is not everybody, is to contribute his part to the promotion of human happiness in the order stated. No doubt the definition goes more completely to the root of the matter than the German formula. It does not trouble itself with the harmony, the many-sidedness, the wholeness, of the individual development; it would admit these just as might be requisite for securing the final end.

James Mill is not singular in his over-grasping view of the subject. The most usual subdivision of Education is into Physical, Intellectual, Moral, Religious, Technical. Now when we inquire into the meaning of Physical Education, we find it to be the rearing of a healthy human being, by all the arts and devices of nursing, feeding, clothing, and general regimen. Mill includes this subject in his article, and Mr. Herbert Spencer devotes a very interesting chapter to it in his work on Education. It seems to me, however, that this department may be kept quite separate, important though it be. It does not at all depend upon the principles and considerations that the educator, properly so called, has in view in the carrying on of his work. The discussion of the subject does not in any way help us in educational matters, as most commonly understood; nor does it derive any illumination from being placed side by side with the arts of the recognized teacher. The fact of bodily health or vigour is a leading postulate in bodily or mental training, but the trainer does not take upon himself to lay down the rules of hygiene.

The inadvertence—for so I regard it—of coupling the Art of Health with Education is easily disposed of, and does not land us in any arduous controversies. Very different is another aspect of these definitions : that wherein the end of Education is propounded as the promotion of human happiness, human virtue, human perfection. Probably the qualification will at once be conceded, that Education is but one of the means, a single contributing agency to the all-including end. Nevertheless, the openings for difference of opinion as to what constitutes happiness, virtue, or perfection are very wide. Moreover, the discussion has its proper place in Ethics and in Theology, and if brought into the field of Education, should be received under protest.

Before entering upon the consideration of this difficulty, the greatest of all, I will advert to some of the other views of Education that seem to err on the side of including too much. Here I may quote from the younger Mill, who, like his father, and unlike the generality of theorists, starts *more scientifico* with a definition. Education, according to him, 'includes whatever we do for ourselves, and whatever is done for us by others, for the express purpose of bringing us nearer to the perfection of our nature ; in its largest acceptation it comprehends even the indirect effects produced on character and on the human faculties by things of which the direct purposes are different ; by laws, by forms of government, by the industrial arts, by modes

of social life ; nay, even by physical facts not dependent
on human will ; by climate, soil, and local position.' He
admits, however, that this is a very wide view of the
subject, and for his own immediate purpose advances a
narrower view, namely—' the culture which each genera-
tion purposely gives to those who are to be its successors,
in order to qualify them for at least keeping up, and, if
possible, for raising, the improvement which has been
attained.' [1]

Besides involving the dispute as to what constitutes
' perfection,' the first and larger statement is, I think,
too wide even for the most comprehensive Philosophy of
Education. The influences exerted on the human character by climate and geographical position, by arts,
laws, government, and modes of social life, constitute a
very interesting department of Sociology, and have their
place there and nowhere else. What we do for ourselves,
and what others do for us, to bring us nearer to the perfection of our nature, may be education in a precise sense
of the word, and it may not. I do not see the propriety
of including under the subject the direct operation of
rewards and punishments. No doubt we do something
to educate ourselves, and society does something to
educate us, in a sufficiently proper acceptation of the
word ; but the ordinary influence of society, in the dispensing of punishment and reward, is not the essential
fact of Education, as I propose to regard it, although an
adjunct to some of its legitimate functions.

Mill's narrower expression of the scope of the subject
is not exactly erroneous ; the moulding of each genera-
tion by the one preceding is not improperly described

[1] *Inaugural Address at St. Andrews*, p, 4.

as an education. It is, however, grandiose rather than scientific. Nothing is to be got out of it. It does not give the lead to the subsequent exposition.

I find in the article 'Education,' in 'Chambers's Encyclopædia,' a definition to the following effect :—' In the widest sense of the word a man is *educated*, either for good or for evil, by everything that he experiences from the cradle to the grave [say, rather, "formed," "made," "influenced"]. But in the more limited and usual sense, the term education is confined to the efforts made, of set purpose, to train men in a particular way —the efforts of the grown-up part of the community to inform the intellect and mould the character of the young [rather too much stress on the fact of influence from without] ; and more especially to the labours of professional educators or schoolmasters.' The concluding clause is the nearest to the point—the arts and methods employed by the schoolmaster ; for, although he is not alone in the work that he is expressly devoted to, yet he it is that typifies the process in its greatest singleness and purity. If by any investigations, inventions, or discussions we can improve his art to the ideal pitch, we shall have done nearly all that can be required of a science and art of Education.

I return to the greater difficulty—namely, the question what is the end of all teaching ; or, if the end be human happiness and perfection, what definite guidance does this furnish to the educator ? I have already remarked that the inquiry is acknowledged to belong to other departments ; and, if in these departments clear and unanimous answers have not been arrived at, the educationist is not bound to make good the deficiency.

For this emergency, there is one thing obvious, another less obvious; the two together exhausting the resources of the educator.

The obvious thing is to fix upon whatever matters people are agreed·upon. Of such the number is considerable, and the instances important. They make the universal topics of the schools.

The less obvious thing is, with reference to matters not agreed upon, that the educator should set forth at what cost these doubtful acquisitions would have to be made ; for the cost must be at least one element in the decision respecting them. Whoever knows most about Education is best able to say how far its appliances can cope with such aims as softening the manners, securing self-renunciation, bringing about the balanced action of all the powers, training the whole man, and so forth.

We shall see that one part of the science of Education consists in giving the ultimate analysis of all complex growths. It is on such an analysis that the cost can be·calculated ; and, by means of this, we can best observe whether contradictory demands are made upon the educator.

What we have been drifting to, in our search for an aim, is the work of the school. This may want a little more paring and rounding to give it scientific form, but it is the thing most calculated to fix and steady our vision at the outset.

Now, for the success of the schoolmaster's work, the first and central fact is the plastic property of the mind itself. On this depends the acquisition not simply of knowledge but of everything that can be called an ac-

3

quisition. The most patent display of the property consists in memory for knowledge imparted. In this view
the leading inquiry in the art of Education is how to
strengthen memory. We are therefore led to take
account of the several mental aptitudes that either
directly or indirectly enter into the retentive function.
In other words, we must draw upon the science of the
human mind for whatever that science contains respecting the conditions of memory.

Although memory, acquisition, retentiveness, depends
mainly upon one unique property of the intellect, which
accordingly demands to be scrutinised with the utmost
care, there are various other properties, intellectual and
emotional, that aid in the general result, and to each of
these regard must be had, in a Science of Education.

We have thus obtained the clue to one prime division
of the subject—the purely psychological part. Of no
less consequence is another department at present without a name—an inquiry into the proper or natural order
of the different subjects, grounded on their relative
simplicity or complexity, and their mutual dependence.
It is necessary to success in Education that a subject
should not be presented to the pupil until all the preparatory subjects have been mastered. This is obvious
enough in certain cases : arithmetic is taken before
algebra, geometry before trigonometry, inorganic chemistry before organic ; but in many cases the proper
order is obscured by circumstances, and is an affair of
very delicate consideration. I may call this the Analytic
or Logical branch of the theory of Education.

It is a part of scientific method to take strict account
of *leading terms*, by a thorough and exhaustive inquiry

into the meanings of all such. The settlement of many questions relating to education is embarrassed by the vagueness of the single term ' discipline.'

Further, it ought to be pointed out, as specially applicable to our present subject, that the best attainable knowledge on anything is due to a combination of general principles obtained from the sciences, with well-conducted observations and experiments made in actual practice. On every great question there should be a convergence of both lights. The technical expression for this is 'the union of the Deductive and Inductive Methods.' The deductions are to be obtained apart, in their own way, and with all attainable precision. The inductions are the maxims of practice—purified, in the first instance, by wide comparison and by the requisite precautions.

I thus propose to remove from the Science of Education matters belonging to much wider departments of human conduct, and to concentrate the view upon what exclusively pertains to Education—the means of building up the acquired powers of human beings. The communication of knowledge is the ready type of the process, but the training operation enters into parts of the mind not intellectual—the activities and the emotions ; the same forces, however, being at work.

Education does not embrace the employment of *all* our intellectual functions. There is a different art for directing the faculties in productive labour ; as, for example, in the professions, in the original investigations of the man of science, in the creations of the artist. The principles of the human mind are applicable to both departments, but although the two come into occasional contact, they are so far distinct that there is an advantage in viewing them separately.   In the practical treatise of Locke, entitled 'The Conduct of the Understanding,' acquisition, production, and invention are handled promiscuously.

CHAPTER II.

THE science of Physiology, coupled with the accumulated empirical observations of past ages, is the theoretical guide in finding out how to rear living beings to the full maturity of their physical powers. This, as we have said, is quite distinct from the process of Education.

The art of Education assumes a certain average physical health, and does not inquire into the means of keeping up or increasing that average. Its point of contact with physiology and hygiene is narrowed to the plastic or acquisitive function of the brain—the property of cementing the nervous connections that underlie memory, habit, and acquired power.

But as Physiology now stands, we soon come to the end of its applications to the husbanding of the plastic faculty. The inquiry must proceed upon our direct experience in the work of education, with an occasional check or caution from the established physiological laws. Still, it would be a forgetting of mercies to undervalue the results accruing to education from the physiological doctrine of the physical basis of memory.

On this subject, physiology teaches the general fact

that memory reposes upon a nervous property or power, sustained, like every other physical power, by nutrition, and having its alternations of exercise and rest. It also informs us that, like every other function, the plasticity may be stunted by inaction, and impaired by over-exertion.

As far as pure physiology is concerned, I would draw attention to one circumstance in particular.   The human body is a great aggregate of organs or interests—digestion, respiration, muscles, senses, brain.   When fatigue overtakes it, the organs generally suffer ; when renovation has set in, the organs generally are invigorated. This is the first and most obvious consequence. We have next to add the qualifying consideration that human beings are unequally constituted as regards the various functions ; some being strong in stomach, others in muscle, others in brain. In all such persons the general invigoration is equally shown; the favoured organs receive a share proportioned to their respective capitals : to him that hath shall be given.   Still more pertinent is the further qualification, that the organ that happens to be most active at the time receives more than its share ; to exercise the several organs unequally is to nourish them unequally.

Now comes the important point. To increase the plastic property of the mind, you must nourish the brain. You naturally expect that this result will ensue when the body generally is nourished : and so it will, if there be no exorbitant demands on the part of other organs, giving them such a preference as to leave very little for the organ of the mind. If the muscles or the digestion are unduly drawn upon, the brain will not respond to

the drafts made upon it. Obversely, if the brain is constituted by nature, or excited by stimulation, so as to
absorb the lion's share of the nutriment, the opposite
results will appear ; the mental functions will be exalted,
and the other interests more or less impoverished. This
is the situation for an abundant display of mental
force.

But we must further distinguish the mental functions
themselves ; for these are very different and mutually
exclusive. Great refinement in the subdivisions is not
necessary for the illustration. The broadest contrast is
the emotional and the intellectual—feeling as pleasure,
pain, or excitement, and feeling as knowledge. These
two in extreme manifestation are hostile to each other :
under excessive emotional excitement the intellect
suffers ; under great intellectual exertion the emotions
subside (with limitations unnecessary for our purpose).

But Intellect in the largest sense is not identical with
the retentive or plastic operation. The laws of this
peculiar phase of our intelligence are best obtained by
studying it as a purely mental fact. Yet there is a
physiological way of looking at it that is strongly con-
firmative of our psychological observations. On the
physical or physiological side, memory or acquisition is
a series of new nervous growths, the establishment of a
number of beaten tracks in certain lines of the cerebral
substance. Now, the presumption is, that as regards the
claim for nourishment this is the most costly of all the
processes of the intelligence. To exercise a power once
acquired should be a far easier thing, much less expensive, than to build up a new acquirement. We may be
in sufficiently good condition for the one, while wholly

out of condition for the other. Indeed, success in acquirement, looking at it according to the physiological probabilities, should be the work of rare, choice, and happy moments: times when cerebral vigour is both abundant and well-directed.

CHAPTER III.

THE largest chapter in the Science of Education must be the following out of all the psychological laws that bear directly or indirectly upon the process of mental acquirement. Every branch of Psychology will be found available; but more especially the Psychology of the Intellect. Of the three great functions of the Intellect, in the ultimate analysis—Discrimination, Agreement, Retentiveness—the last is the most completely identified with the educative process; but the others enter in as constituents in a way peculiar to each.

DISCRIMINATION.

Mind starts from Discrimination. The consciousness of difference is the beginning of every intellectual exercise. To encounter a new impression is to be aware of change: if the heat of a room increases ten degrees, we are awakened to the circumstance by a change of feeling; if we have no change of feeling, no altered consciousness, the outward fact is lost upon us: we take no notice of it, we are said not to know it.

Our intelligence is, therefore, absolutely limited by our power of discrimination. The other functions of intellect, the Retentive power, for example, are not called into play, until we have first discriminated a number of things. If we did not originally feel the difference between light and dark, black and white, red and yellow, there would be no visible scenes for us to remember: with the amplest endowment of Retentiveness, the outer world could not enter into our recollection ; the blank of sensation is a blank of memory.

Yet further. The minuteness or delicacy of the feeling of difference is the measure of the variety and multitude of our primary impressions, and, therefore, of our stored-up recollections. He that hears only twelve discriminated notes on the musical scale, has his remembrance of sounds bounded by these ; he that feels a hundred sensible differences, has his ideas or recollections of sounds multiplied in the same proportion. · The retentive power works up to the height of the discriminative power ; it can do no more.

We have by nature a certain power of discrimination in each department of our sensibility. We can from the outset discriminate, more or less delicately, sights, sounds, touches, smells, tastes ; and, in each sense, some persons much more than others. This is the deepest foundation of disparity of intellectual character, as well as of variety in likings and pursuits. If, from the beginning, one man can interpolate five shades of discrimination of colour where another can feel but one transition, the careers of the two men are foreshadowed and will be widely apart.

To observe this native inequality is important in predestining the child to this or that line of special training. For the actual work of teaching, it is of more consequence to note the ways and means of quickening and

Increasing the discriminating aptitude. Bearing in mind the fact that, until a difference is felt between two things, intelligence has not yet made the first step, the teacher is bound to consider the circumstances or conditions favourable and unfavourable to the exercise.

(1) It is not peculiar to discrimination, but is common to every intellectual function, to lay down, as a first condition, mental vigour, freshness, and wakefulness. In a low state of the mental forces, in languor, or drowsiness, differences cannot be felt. That the mind should be alive, awake, in full force and exercise, is necessary for every kind of mental work. The teacher needs to quicken the mental alertness by artificial means when there is a dormancy of mere indolence. He has to waken the pupil from the state significantly named *indifference*, the state where differing impressions fail to be recognised as distinct.

(2) The mind may be fresh and alive, but its energies may be taking the wrong direction. There is a well-known antithesis or opposition between the emotional and the intellectual activities, leading to a certain incompatibility of the two. Under emotional excitement, the intellectual energies are enfeebled in amount, and enslaved to the reigning emotion. It is in the quieter states of mind that discrimination, in common with other intellectual powers, works to advantage. I will afterwards discuss more minutely the very delicate matter of the management of the various emotions in the work of teaching.

(3) It must not be forgotten that intellectual exercises are in themselves essentially insipid, unattractive. As exertion, they impart a certain small degree of the

delight that always attends the healthy action of an exuberant faculty ; but this supposes their later developments, and is not a marked peculiarity in the child's commencing career. The first circumstance that gives an interest to discrimination is pleasurable or painful stimulus. Something must hang on a difference before the mind is made energetically awake to it. A thoroughly uninteresting difference is not an object of attention to anyone,

The transitions from cold to hot. dark to light, strain to relief, hunger to repletion, silence to sound, are all more or less interesting, and more or less impressive. But then they are vehement and sensational. It is necessary, in order to the furnishing of the intelligence, that smaller and less sensational transitions should be felt ; the intellectual nature is characterized by requiring the least amount of emotional flash in order to impress a difference. A loud and furious demonstration will certainly compel attention and end in the feeling of difference, but at too much cost.

(4) The great practical aid to the discovery and the retention of difference is immediate succession, or, what comes to the same thing, close juxtaposition. A rapid transition makes evident a difference that would not be felt after an interval, still less if anything else were allowed to occupy the mind in the meantime. This fact is sufficiently obvious, and is turned to account in easy cases ; but is far from being thoroughly worked out by the teacher and the expositor. Any trifling diversion will suffice to blind us to its importance.

We compare two notes by sounding them in close succession ; two shades of colour by placing them side by side ; two weights by holding them in the two hands, and attending to the two feelings by turns. These are the plain instances. The comparison of forms leads to complications, and we cease to attempt the same kind of comparison. For mere length we lay the two things alongside ; so for an angle. For number, we can place two groups in contiguous rows—three by the side of four or five--and observe the surplus.

Mere size is an affair of simple juxtaposition. Form, irrespective of size, is less approachable. A triangle and a quadrangle are compared by counting the sides, and resolving the difference of form into the simpler element of difference of number. A right-angled, an acute-angled, an isosceles triangle, must be compared by the juxtaposition of angles. A circle and an oval are contrasted by the curvature and the diameters ; in the one, the curvature uniform. and the diameters equal ; in the other, the curvature varying, and the diameters unequal. The difference between a close and an open curve is palpable enough.

The geometrical forms are thus resolvable into very simple bases of comparison ; and the teacher must analyze them in the manner now stated. For the irregular and capricious forms, the elementary conceptions are still the same—lineal size, number, angular size, curvature ; but the mode of guiding the attention may be various. Sometimes there is a strong and overpowering similarity, with a small and unconspicuous difference, as in our ciphers (compare 3 and 5), and in the letters of our alphabet (C, G), and still more in the Hebrew alphabet. For such comparisons, the difference, such as it is, needs to be very clearly drawn or even exaggerated.

Another method is to have models of the same size to lay over one another, so as to bring out the difference through juxtaposition. By a specific effort, the teacher calls on the learner to view, with single-minded attention, the differing circumstance, and afterwards to reproduce it by his own hand. One express lesson consists in asking the pupil what are the ciphers, or the letters, that are nearly alike, and what are the points of difference.

The higher arts of comparison to impress difference are best illustrated when both differences and agreements have to be noted. They will have to be resumed after the discussion of the intellectual force of Agreement or Similarity. The chief stress of the present explanation lies in regarding Discrimination as the necessary prelude of every intellectual impression, as the basis of our stored-up knowledge, or memory. Agreement is presupposed likewise ; but there is not the same necessity, nor is it expedient, to follow out the workings of Agreement, before considering the plastic power of the intellect.

THE RETENTIVE FACULTY.

This is the faculty that most of all concerns us in the work of Education. On it rests the possibility of mental growths ; in other words, capabilities not given by nature.

Every impression made upon us, if sufficient to awaken consciousness at the time, has a certain permanence ; it can persist after the original ceases to work ; and it can be restored afterwards as an idea or remembered impression. The bursting out of a flame arouses oui attention, gives a strong visible impression, and becomes an idea or deposit of memory. The flame is thought of afterwards without being actually seen.

It is not often that one single occurrence leaves a permanent and recoverable idea ; usually, we need several repetitions for the purpose. The process of fixing the impression occupies a certain length of time ; either we must prolong the first shock, or renew it on several successive occasions. This is the first law of Memory, Retention, or Acquisition : ' Practice makes perfection ;' ' Exercise is the means of strengthening a faculty ;' and so forth. The good old rule of the schoolmaster is simply to make the pupil repeat, rehearse, or persist at, a lesson, until it is learnt.

All improvement in the art of teaching depends on the attention that we give to the various circumstances that facilitate acquirement, or lessen the number of repetitions for a given effect. Much is possible in the way of economizing the plastic power of the human system ; and when we have pushed this economy to the utmost, we have made perfect the Art of Education in one leading department. It is thus necessary that the consideration of all the known conditions that favour or impede the plastic growth of the system should be searching and minute

Although some philosophers have taught that all minds are nearly equal in regard to facility of acquirement, a schoolmaster that would say so must be of the very rudest type. The inequality of different minds in imbibing lessons, under the very same circumstances, is a glaring fact ; and is one of the obstacles encountered in teaching numbers together, that is, classes. It is a difficulty that needs a great deal of practical tact or management, and is not met by any educational theory.

The different kinds of acquirements vary in minor circumstances, which call for notice after we have exhausted the general or pervading conditions. The greatest contrast is between what belongs to Intelligence, and what belongs to the Feelings and the Will. The more strictly Intellectual department comprises Mechanical Art, Language, the Sensible World, the Sciences. Fine Art ; each having their specialities.

General Circumstances Favouring Retentiveness.

1. The Physical condition. This has been already touched upon, both in the review of Physiology, and in the remarks on Discrimination. It includes general health, vigour, and freshness at the moment, together with the further indispensable proviso, that the nutrition, instead of being drafted off to strengthen the mere physical functions, is allowed to run in good measure to the brain.

In the view of mental efficiency, the muscular system, the digestive system, and the various organic interests, are to be exercised up to the point that conduces to the maximum of general vigour in the system, and no farther. They may be carried farther in the interest of sensual enjoyment, but that is not now before us. Hence a man must exercise his muscles, must feed himself liberally, and give time to digestion to do its work, must rest adequately—all for the greatest energy of the mind, and for the trying work of education in particular. Nor is it so very difficult, in the present state of physiological and medical knowledge, to assign the reasonable proportions in all these matters, for a given case.

Everything tends to show that, in the mere physical point of view, the making of impressions on the brain, although never remitted during any of our waking moments, is exceedingly unequal at different times. We must be well aware that there are moments when we are incapable of receiving any lasting impressions, and that there are moments when we are unusually susceptible. The difference is not one wholly resolvable into more mental energy on the whole ; we may have a considerable reserve of force for other mental acts, as the performance of routine offices, and not much for retaining new impressions ; we are capable of reading, talking, writing, and of taking an interest in the exercises ; we may indulge emotions, and carry out pursuits, and yet not be in a state for storing the memory, or amassing knowledge. Even the incidents that we take part in sometimes fail to be remembered beyond a very short time.

What, then, is there so very remarkable and unique in the physical support of the plastic property of the brain ? What are the moments when it is at the plenitude of its efficiency ? What are the things that especially nourish and conserve it ?

Although there is still wanting a careful study of this whole subject, the patent facts appear to justify us in asserting, that the plastic or retentive function is the *very highest energy* of the brain, the consummation of nervous activity. To drive home a new bent, to render an impression self-sustaining and recoverable, uses up (we may suppose) more brain force than any other kind of mental exercise. The moments of susceptibility to

4

the storing up of knowledge, to the engraining of habits and acquisitions, are thus the moments of the maximum of unexpended force. The circumstances need to be such as to prepare the way for the highest manifestation of cerebral energy; including the perfect freshness of the system, and the absence of everything that would " speedily impair it.

To illustrate this position, I may refer to the kind of mental work that appears to be second in its demand on the energy of the brain. The exercise of mental con- structiveness—the solving of new problems, the applying of rules to new cases, the intellectual labour of the more arduous professions, as the law—demands no little men- tal strain, and is easy according to the brain vigour of the moment. Still, these are exercises that can be per- formed with lower degrees of power; we are capable of such professional work in moments when our memory would not take in new and lasting impressions. In old age, when we cease to be educable in any fresh endow- ment, we can still perform these constructive exercises ; we can grapple with new questions, invent new argu- ments and illustrations, decide what should be done in original emergencies.

The constructive energy has all degrees, from the highest flights of invention and imagination down to the point where construction shades off into literal repetition of what has formerly been done. The preacher in com- posing a fresh discourse puts forth more or less of con- structiveness: in repeating prayers and formularies, in reading from book, there is only reminiscence. This last is the third and least exigent form of mental energy ; it is possible in the very lowest states of cerebral vigour.

When acquisition is fruitless, construction is possible; when a slight departure from the old routine passes the might of the intelligence, literal reminiscence may operate.

Another mode of mental energy that we are equal to, when the freshness of our susceptibility to new growths has gone off, is searching and noting. This needs a certain strain of attention; it is not possible in the very lowest tide of the nervous flow; but it may be carried on with all but the smallest degrees of brain power. When the scholar or the man of science ceases to trust his memory implicitly for retaining new facts that occur in his reading, observation, or reflection, he can still keep a watch for them, and enter them in his notes. So in the hours of the day when memory is less to be trusted, useful study may still be maintained by the help of the memorandum and the note-book.

The indulgence of the emotions (when not violent or excessive) is about the least expensive of our mental exercises, and may go on when we are unfit for any of the higher intellectual moods, least of all for the crowning work of storing up new knowledge or new aptitudes. There are degrees here also; but, speaking generally, to love or to hate, to dominate or to worship, although impossible in the lowest depths of debility, are within the scope of the inferior grades of nervous power.

From this estimate of comparative outlay, we may judge what are the times and seasons and circumstances most favourable to the work of acquirement. It may be assumed that in the early part of the day the total energy of the system is at its height, and that towards evening it flags; hence morning is the season of improvement.   For two or three hours after the first meal, the strength is probably at the highest; total remission for another hour or two, and a second meal (with physical exercise when the labour has been sedentary), prepare for a second display of vigour, although presumably not equal to the first, except in youthful years; when the edge of this is worn off, there may, after a pause, be another bout of application, but far inferior in result to the first or even to the second.   No severe effort should be attempted in this last stage; not much stress should be placed on the available plasticity of the system, although the constructive and routine efforts may still be kept up.

The regular course of the day may be interfered with by exceptional circumstances, but these only confirm the rule.   If we have lain idle or inactive for the early hours, we may, of course, be fresher in the evening, but the late application will not make up for the loss of the early hours; the nervous energy will gradually subside as the day advances, however little exertion we may make.   Again, we may at any time determine an outburst of nervous energy by persistent exercise and by stimulation, which draws blood to the brain, without regard to circumstances and seasons, but this is wasteful in itself and disturbing to the healthy functions.

As a general rule, the system is at its greatest vigour in the cold season of the year; and most work is done in winter.   Summer studies are comparatively unproductive.

The review of the varying plasticity in the different stages of life might be conducted on the same plan of estimating the collective forces of the system, and the share of these available for brain work, but other circumstances have to be taken into the account, and I do not enter upon the question here.

There are many details in the economy of the plastic power that have a physical as well as a mental aspect. Such are those relating to the strain and remission of the Attention, to the pauses and alternations during the times of drill, to the moderating of the nervous excitement, and other matters. These should all find a place under the head of the Retentive function. It is expedient now to take up the consideration of the subject from the purely mental side.

2. The one circumstance that sums up all the mental aids to plasticity is CONCENTRATION. A certain expenditure of nervous power is involved in every adhesion, every act of impressing the memory, every communicated bias ; and the more the better. This supposes, however, that we should withdraw the forces, for the time, from every other competing exercise ; and especially, that we should redeem all wasting expenditure for the purpose in view.

It is requisite, therefore, that the circumstances leading to the concentration of the mind should be well understood. We assume that there is power available for the occasion, and we seek to turn it into the proper channel. Now, there is no doubt that the will is the chief intervening influence, and the chief stimulants of the will are, as we know, pleasure and pain. This is the rough view of the case. A little more precision is attainable through our psychological knowledge.

And first, the Will itself as an operating or directing power, that is tc say, the moving of the organs in a given way under a motive, is a growth or culture; it is very imperfect at first, and improves by usage. A child of twelve months cannot by any inducement be prompted readily to clap its hands, to point with its forefinger, to touch the tip of its nose, to move its left shoulder forward. The most elementary acts of the will, the alphabet of all the higher acquisitions, have first to be learned in a way of their own; and until they have attained a sufficient advancement, so as to be amenable to the spur of a motive, the teacher has nothing to go upon.

I have elsewhere described this early process, as I conceive it, in giving an account of the development of the Will. In the practice of education, it is a matter of importance as showing at what time mechanical instruction is possible, and what impedes its progress at the outset, notwithstanding the abundance of plasticity in the brain itself. The disciplining of the organs to follow directions would seem to be the proper province of the Infant school.

Coming now to the influences of concentration, we assign the first place to intrinsic charm, or *pleasure in the act* itself. The law of the Will, on its side of greatest potency, is—that Pleasure sustains the movement that brings it. The whole force of the mind at the moment goes with the pleasure-giving exercise. The harvest of immediate pleasure stimulates our most intense exertions, if exertion serves to prolong the blessing. So it is with the deepening of an impression, the confirming of a bent or bias, the associating of a couple or a sequence of acts; a coinciding burst of joy awakens the attention, and thus leads to an enduring stamp on the mental framework.

The engraining efficiency of the pleasurable motive requires not only that we should not be carried off into an accustomed routine of voluntary activities, such as to give to the forces another direction, as when we pace to and fro in a flower garden ; but also that the pleasure should not be intense and tumultuous.   The law of the mutual exclusion of great pleasure and great intellectual exertion forbids the employment of too much excitement of any kind, when we aim at the most exacting of all mental results—the forming of new adhesive growths. A gentle pleasure that for the time contents us, there being no great temptation at hand, is the best foster-mother of our efforts at learning.   Still better, if it be a growing pleasure ; *a small beginning, with steady increase*, never too absorbing, is the best of all stimulants to mental power.   In order to have a yet wider compass of stimulation, without objectionable extremes, we might begin on the negative side, that is, in pain or privation, to be gradually remitted in the course of the studious exercise, giving place at last to the exhilaration of a waxing pleasure.   All the great teachers, from Socrates downwards, seem to recognize the necessity of putting the learner into a state of pain to begin with ; a fact that we are by no means to exult over, although we may have to admit the stern truth that is in it. The influence of pain, however, takes a wider range than is here supposed, as will be seen under our next head.

A moderate exhilaration and cheerfulness growing out of the act of learning itself is certainly the most genial, the most effectual means of cementing the unions that we desire to form in the mind. This is meant when we speak of the learner having a taste for his pursuit, having the *heart* in it, learning *con amore*. The fact is perfectly well known; the error, in connection with it, lies in dictating or enjoining this state of mind on every-body in every situation, as if it could be commanded by a wish, or as if it were not itself an expensive endow-ment. The brain cannot yield an exceptional pleasure without charging for it.

Next to pleasure in the actual, as a concentrating motive, is pleasure in *prospect*, the learning of what is to bring us some future gratification. The stimulus has the inferiority attaching to the idea of pleasure as compared with the reality. Still it may be of various degrees, and may rise to a considerable pitch of force. Parents often reward their children with coins for success in their lessons; the conception of the pleasure in this case is nearly equal to a present tremor of sense-delight. On the other hand, the promises of fortune and distinction, after a long interval of years, have seldom much influ-ence in concentrating the mind towards a particular study.

Let us now view the operation of Pain. By the law of the will, pain makes us recoil from the thing that causes it. A painful study repels us, just as an agreeable one attracts and detains us. The only way that pain can operate is when it is attached to neglect, or to departure from the prescribed subject; we then find pleasure, by comparison, in sticking to our task. This is the theory of punishing the want of application. It is in every way inferior to the other motives; and this inferiority should be always kept in view in employing it, as indeed every teacher must often do with the generality of scholars. Pain is a waste of brain-power; while the work of the learner needs the very highest form of this power. Punishment works at a heavy percentage of deduction, which is still greater as it passes into the well-defined form of terror. Everyone has experienced cases where severity has rendered a pupil utterly incapable of the work prescribed.

Discarding all *à priori* theories as to whether the human mind can be led on to study by an ingenious system of pleasurable attractions, we are safe to affirm that if the physical conditions are properly regarded, if the work is within the compass of the pupil's faculties, and if a fair amount of assistance is rendered in the way of intelligible direction, although some sort of pain will frequently be necessary, it ought not to be so great as to damp the spirits and waste the plastic energy.

The line of remark is exactly the same for pain in prospect, with allowance for the difference between reality and the idea. It is well when prospective pain has the power of a motive, because the future bad consequences of neglect are so various and so considerable, as to save the resort to any other. But since the young intelligence in general is weak in the sense of futurity, whether for good or for evil, none but very near, very intelligible, and very certain pains can take the place of presently acting deterrents.

In the study of the human mind, we need, for many purposes, to draw a subtle distinction between feeling as Pleasure or Pain, and feeling as Excitement not necessarily pleasurable or painful. This subtlety cannot be dispensed with in our present subject. There is a form of mental concentration that is properly termed excitement, and is not properly termed pleasurable or painful excitement. A loud or sudden shock, a rapid whirling movement, stirs, wakens, or excites us ; it may also give us pleasure or pain, but it may be perfectly neutral; and even when there is pleasure or pain, there is an influence apart from what would belong to pleasure or pain, as such. A state of excitement seizes hold of the mind for the time being and shuts out other mental occupations ; we are engrossed with the subject that brought on the state, and are not amenable to extraneous influences, until that has subsided. Hence, excitement is pre-eminently a means of making an impression, of stamping an idea in the mind : it is strictly an intellectual stimulus. There is still the proviso (under the general law of incompatibility of the two opposite moods) that the excitement must not be violent and wasting. In well-understood moderation, excitement is identical with attention, mental engrossment, the concentration of the forces upon the plastic or cementing operation, the rendering permanent as a recollection what lies in the focus of the blaze. Excitement, so defined, is worthless as an end, but is valuable as a means ; and that means is the furtherance of our mental improvement by driving home some useful concatenation of ideas.

Another subtlety remains—a distinction within a distinction. After contrasting feeling as excitement with feeling as pleasure or pain, we must separate the useful from the useless or even pernicious modes of excitement. The useful excitement is what is narrowed and confined to the subject to be impressed ; the useless, and worse than useless, excitement is what spreads far and wide, and embraces- nothing in particular. It is easy to get up the last-named species of excitement—

the vague, scattered, and tumultuous mode—but this is not of avail for any set purpose; it may be counted rather as a distracting agency than as a means of calling forth and concentrating the attention upon an exercise.

The true excitement for the purpose in view is what grows out of the very subject itself, embracing and adhering to that subject. Now, for this kind of excitement, the recipe is continuous application of the mind in perfect outside stillness. Restrain all other solicitation of the senses, keep the attention upon the one act to be learnt; and, by the law of nervous and mental persistence, the currents of the brain will become gradually stronger and stronger, until they have reached the point when they do no more good for the time. This is the ideal of concentration by neutral excitement.

The enemy of such happy neutrality is pleasure from without; and the youthful mind cannot resist the distraction of a present pleasure, or even the scent of a far-off pleasure. The schoolroom is purposely screened from the view of what is going on outside; while all internal incidents that hold out pleasurable diversion are carefully restrained, at least during the crisis of a difficult lesson. A touch of pain, or the apprehension of it, *if only slight*, is not unfavourable to the concentration.

An important point is still to be observed, namely, that relationship of Retention to Discrimination that was stated in introducing the function of Discrimination The consideration of this relationship illustrates with still greater point the true character of the excitement that concentrates and does not either distract or dissipate the energies. The moment of a delicate discrimination is the moment when the intellectual force is dominant;

emotion spurns nice distinctions, and incapacitates the mind for feeling them. The quiescence and stillness of the emotions enables the mind to give its full energies to the intellectual processes generally; and of these, the fundamental is perception of difference. Now, the more mental force we can throw into the act of noting a differ- ence, the better is that difference felt, and *the better it is impressed.* The same act that favours discrimination, favours retention. The two cannot be kept separate. No law of the intellect appears to be more certain than the law that connects our discriminating power with our retentive power. In whatever class of subjects our dis- crimination is great—colours, forms, tones, tastes—in that class our retention is great. Whenever the atten- tion can be concentrated on a subject in such a way as to make us feel all its delicate lineaments, which is another way of stating the sense of differences, through that very circumstance a great impression is made on the memory; there is no more favourable moment for engraining a recollection.

The perfection of neutral excitement, therefore, is typified by the intense rousing of the forces in an act, or a series of acts, of discrimination. If by any means we can succeed in this, we are sure that the other intellec- tual consequences will follow. It is a rare and difficult attainment in childhood and early youth : the conditions, positive and negative, for its highest consummation can- not readily be had. Yet we should know what these conditions are ; and the foregoing attempt has been made to seize and embody them.

Pleasure and pain, besides operating in their own character, that is, besides directing the voluntary actions, have a power as mere excitement, or as wakening up the mental blaze, during which all mental acts, including the impressing of the memory, are more effective. The distinction must still be drawn between concentrated and diffused excitement, between excitement *in*, and excitement *away from*, the work to be done. Pleasure is the more favourable adjunct, if not too great. Pain is the more stimulating or exciting : under a painful smart the forces are very rapidly quickened for all purposes, until we reach the point of wasteful dissipation. This brings us round again to the Socratic position, the preparing of the learner's mind by the torpedo or the gad-fly.

The full compass of the operation of the painful stimulant is well shown in some of our most familiar experiences as learners. In committing a lesson to memory, we con it a number of times by the book : we then try without the book. We fail utterly, and are slightly pained by the failure. We go back to the book, and once more we try without it. We still fail, but rack the memory to recover the lost trains. The pains of failure and the act of straining stimulate the forces ; the attention is roused seriously and energetically. The next reference to the book finds us far more receptive of the impression to be made ; the weak links are now reinforced with avidity, and the next trial shows the value of the discipline that has been undergone.

One remark more will close the view of the conditions of plasticity. It is, that Discrimination and Retentiveness have a common support in rapidity and sharpness of transition. A sharp and sudden change is commonly said to make a strong *impression* : the fact implied concerns discrimination and retention alike.

Vague, shadowy, ill-defined boundaries fail to be dis·
criminated, and the subjects of them are not remembered.
The educator finds great scope for his art in this consideration also.

SIMILARITY, OR AGREEMENT.

It is neither an inapt nor a strained comparison to
call this power the Law of Gravitation of the intellectual
world. As regards the Understanding, it has an importance co-equal with the plastic force that is expressed by
Retentiveness or Memory. The methods to be pursued in attaining the commanding heights of General
Knowledge are framed by the circumstances attending
the detection of Like in the midst of Unlike.

With all the variety that there is in the world of our
experience, a variety appealing to our consciousness of
difference, there is also great Repetition, sameness or
unity. There are many shades of colour, as distinguished by the discriminative sensibility of the eye ; yet
the same shade often recurs. There are many varieties
of form—the round, the square, the spiral, &c.—and we
discriminate them when they are contrasted ; while the
same form starts up again and again. At first sight,
this apparently means nothing at all; the great matter
would seem to be to avoid confounding differences—
blue with violet, a circle with an oval ; when blue recurs,
we simply treat it as we did at first.

The remark is too hasty, and overlooks a vital
consideration. What raises the principle of Similarity
to its position of command is the accompaniment of
diversity. The round form first discerned in a ring or a

halfpenny, recurs in the full moon, where the adjuncts
are totally different and need to be felt as different. In
spite of these disturbing accompaniments, it is important to feel the agreement on the single property called
the round form.

. When an impression made in one situation is repeated in an altered situation, the new experience
reminds us of the old, notwithstanding the diversity ;
this reminder may be described as a novel kind of shock,
or awakened consciousness, called the shock or flash of
identity in the midst of difference. A piece of coal and
a piece of wood differ, and are at first looked upon as
differing. Put into the fire, they both blaze up, give
heat, and are consumed. Here is a shock of agreement,
which becomes an abiding impression in connection
with these two things. Of such shocks is made up one-
half of what we term Knowledge.

Whenever there is a difference it should be felt by
us ; in like manner, whenever there is an agreement it
should be felt. To overlook either one or the other is
stupidity. Our education marches in both lines ; and, in
so far as we are helped by the schoolmaster, we should
be helped in both. The artifices that promote discrimination, and the influences that thwart it, have been
already considered ; and many of the observations apply
also to Agreement. In the identifying of like in the
midst of unlike, there are cases that are easy, and there
are cases that the unassisted mind fails to perceive.

1. We must repeat, with reference to the delicate
perception of Agreements, the *antithesis of the intellectual and the emotional outgoings*. It is in the stillness of the emotions that the higher intellectual exercises

are possible. This circumstance should operate as a warning against the too frequent recourse to pains and penalties, as well as against pleasurable and other excitement. But a more specific application remains.

We may at once face the problem of General Knowledge. The most troublesome half of the education of the intellect is the getting possession of generalities. A general fact, notion, or truth, is a fact recurring under various circumstances or accompaniments. ' Heat' is the name for one such generality. There are many individual facts greatly differing among themselves, but all agreeing in the impression called heat—the sun, a fire, a lamp, a living animal. The intellect discerns, or is struck with, the agreement, notwithstanding the differences; and in this discernment arrives at a general idea.

Now the grand stumbling-block in the way of the generalizing impetus is the presence of the individual differences. These may be small and insignificant, or they may be very great. In comparing fires with one another, the agreement is striking, while the differences between one fire and another, in size, or intensity, or fuel, do not divert the attention from their agreement. But the discerning of sameness in the sun's ray and in a fermenting dung-heap is thwarted by the extraordinary disparity ; and this conflict between the sameness and the difference operates widely and retards the discovery of the most important truths.

2. The device of *juxtaposition* applies to the expounding of Agreement, no less than of Difference. We can arrange the several agreeing facts in such a way that the agreement is more easily seen. The effect is gained partly by closeness, as in the case of differences, and

partly by a symmetrical contact, as when we compare the two hands by placing them finger to finger, and thumb to thumb. Such symmetrical comparisons bring to view, in the same act, agreement and dif. ference. The method reaches far and wide, and is one of the most powerful artificial aids to the im. parting of knowledge

3. The *cumulation* of the instances is essential to the driving home of a generality. A continuous undistracted iteration of the point of agreement is the only way to produce an adequate impression of. a great general idea. I cannot now consider the various obstacles encountered in this attempt, nor explain how seldom it can be adhered to in the highest examples. It must suffice to remark that the interest special to the individual examples is perpetually carrying off the attention ; and pupil and master are both liable to be turned aside by the seduction.

There is another aspect of the power of Similarity, under which it is a valuable aid to Memory or Retention. When we have to learn an exercise absolutely new, we must engrain every step by the plastic adhesiveness of the brain, and must give time and opportunity for the adhesive links to be matured. But when we come to an exercise containing parts already acquired by the plastic operation, we are saved the labour of forging fresh links as regards these, and need only to master what is new to us. When we have known all about one plant, we can easily learn the other plants of the same species or genus ; we need only to master the points of variety.

The bearing of this circumstance on mental growth

5

must be apparent at once. After a certain number of acquirements in the various regions of study—manual art, language, visible pictures—nothing that occurs is absolutely new; the amount of novel matter is continually decreasing as our knowledge increases. Our adhesive faculty is not improving as we grow in years; very much the contrary: but our facility in imbibing new knowledge improves steadily; the fact being that the knowledge is so little new that the forming of fresh adhesions is reduced to a very limited compass. The most original air of music that the most original genius could compose would be very soon learnt by an instructed musician.

In the practice of the schoolmaster's art, this great fact will be perpetually manifesting itself. The operation can be aided and guided in those cases where the agreement really existing is not felt. It is one of the teaching devices, to make the pupils see the old in the new, as far as the agreement reaches; and to pose them upon this very circumstance. The obstacles are the very same as already described, and the means of overcoming them the same. Orderly juxtaposition is requisite for matters of complexity; and we may have also to counterwork the attractions of individuality.[1]

CONSTRUCTIVENESS.

In many parts of our education, the stress lies, not in simple memory, or the tenacious holding of what has

---

[1] When educators prescribe, as a means of impressing the memory, the tracing of the relations of cause and effect, means and end, antecedent and consequent, the appeal is to agreement with some foregone impressions.

been presented to the mind, but in making us perform some new operation, something that we were previously unable to do. Such are the first stages of our instruction in speaking, in writing, and in all the mechanical or manual arts. So also in the higher intellectual processes, as in the imagining of what we have not seen. I do not go so far as to include invention or discovery; the culture of the creative faculty is not comprised in the present discussion.

The psychology of Constructiveness is remarkably simple. There are certain primary conditions that run through all the cases; and it is by paying due respect to these conditions that we can, as teachers, render every possible assistance to the struggling pupils.

1. The constructive process supposes *something to construct from*; some powers already possessed that can be exercised, directed, and combined in a new manner. We must walk before beginning to dance; we must articulate simple sounds before we can articulate words; we must draw straight strokes and pot-hooks before we can form letters; we must conceive trees and shrubs, flowers and grassy plots, before we can conceive a garden.

The practical inference is no less obvious and irresistible; it is one that covers the whole field of education, and can never have been entirely neglected, although it has certainly never been fully carried out. Before entering on a new exercise, we must first be led up to it by mastering the preliminary or preparatory exercises. Teachers are compelled by their failures to attend to this fact in the more palpable exercises, as speaking and writing. They lose sight of it, when the succession of stages is too subtle for their apprehension, as in the understanding of scientific doctrines.

2. In aiming at a new construction, *we must clearly conceive what is aimed at*; we must have the means of judging whether or not our tentatives are successful. The child in writing has the copy lines before it; the man in the ranks sees the fugleman, or hears the approving or disapproving voice of the drill-sergeant. Where we have a very distinct and intelligible model before us, we are in a fair way to succeed; in proportion as the ideal is dim and wavering, we stagger and miscarry. When we depend upon a teacher's expressed approval of our effort, it behoves him to be very consistent, as well as very sound, in his judgment; should he be one thing to-day, and another thing to-morrow, we are unhinged and undone.

It is a defect pertaining to all models that they contain individual peculiarities mixed up with the ideal intention. We carry away with us from every instructor touches of mannerism, and the worst of it is that some learners catch nothing but the mannerism; this being generally easier to fall into than the essential merits of the teaching. There is no remedy here except the comparison of several good models; as the ship-captain carries with him a number of chronometers.

In following an unapproachable original, as in learning to write from copperplate lines, we need a second judgment to inform us whether our deviations are serious and fundamental, or only venial and unavoidable. The good tact of our instructor is here put to the test; he may make our path like the shining light that shineth more and more unto the perfect day, or he may leave us in hopeless perplexity. To point out to us where, how, and why we are wrong, is the teacher's most indispensable function.

3. The only mode of arriving at a new constructive combination is *to try and try again.* The will initiates some movements; these are found not to answer, and are suppressed; others are tried, and so on, until the requisite combination has been struck out. The way to new powers is by trial and error. In proportion as the first and second conditions above given are realized, the unsuccessful trials are fewer. If we have been well led up to the combination required, and if we have before us a very clear idea of what is to be done, we do not need many tentatives; the prompt suppression of the wrong movements ultimately lands us in the right.

The mastering of a new manual combination, as in writing, in learning to swim, in the mechanical arts,—is a very trying moment to the human powers; success involves all those favourable circumstances indicated in discussing the retentive or receptive faculty. Vigour, freshness, freedom from distraction, no strong or extraneous emotions, motives to succeed,—are all most desirable in realizing a difficult combination. Fatigue, fear, flurry, or other wasting excitement, do away with the chances of success.

Very often we have to give up the attempt for a time; yet the ineffectual struggles are not entirely lost. We have at least learnt to avoid a certain number of positions, and have narrowed the round of tentatives for the next occasion. If after two or three repetitions, with rest intervals, the desired combination does not emerge, it is a proof that some preparatory movement

is wanting, and we should be made to retrace the approaches. Perhaps we may have learnt the pre-requisite
movements in a way, but not with sufficient firmness and
certainty for securing their being performed in combination.

ALTERNATION AND REMISSION OF ACTIVITY.

In the accustomed routine of Education, a number
of separate studies and acquirements are prosecuted
together; so that, for each day, a pupil may have to
engage in as many as three, four, or more, different
kinds of lessons.

The principles that guide the alternation and remission of our modes of exercise and application are apparently these :—

1. Sleep is the only entire and absolute cessation
of the mental and bodily expenditure ; and perfect or
dreamless sleep is the greatest cessation of all. Whatever shortens the due allowance of sleep, or renders it
fitful and disturbed, or promotes dreaming, is so much
force wasted.

In the waking hours, there may be cessation from a
given exercise, with more or less of inaction over the
whole system. The greatest diversion of the working
forces is made by our meals : during these the trains of
thought are changed, while the body is rested.

Bodily or muscular exercise, when alternated with
sedentary mental labour, is really a mode of remission
accompanied with an expenditure requisite to redress
the balance of the physical functions. The blood has
unduly flowed to the brain ; muscular exercise draws it
off. The oxidation of the tissues has been retarded ;

muscular exercise is the most direct mode of increasing it. But definite observations teach us that these two beneficial effects are arrested at the fatigue-point; so that the exercise at last contributes not to the refresh· ment, but to the farther exhaustion of the system.

2. The real matter before us is, what do we gain by dropping one form of activity and taking up another? This involves a variety of considerations.

It is clear that the first exercise must not have been pushed so far as to induce general exhaustion. The raw recruit, at the end of his morning drill, is not in a good state to improve his arithmetic in the military school-room. The musical training for the stage is at times so severe as to preclude every other study. The importance of a particular training may be such that we desire for it the whole available plasticity of the system.

It is only another form of exhaustion when the currents of the brain continue in their set channels and refuse any proposed diversion.

There are certain stages in every new and difficult study, wherein it might be well to concentrate for a time the highest energy of the day. Generally, it is at the commencement; but whatever be the point of special difficulty, there might be a remission of all other serious or arduous studies, till this is got over. Not that we need actually to lay aside everything else; but there are, in most studies, many long tracts where we seem in point of form to be moving on, but are really repeating substantially the same familiar efforts. It would be a felicitous ideal adjustment, if the moments of strain in one of the parallel courses were to coincide with the moments of ease in the rest.

Hardly any kind of study or exercise is so complicated and many-sided as to press alike upon all the
energies of the system ; hence there is an obvious propriety in making such variations as would leave unused
as few of our faculties as possible. This principle necessarily applies to every mental process—acquirement,
production, enjoyment. The working out of the principle supposes that we are not led away by the mere
semblance of variety.

Let us endeavour to assign the differences of subject
that afford relief by transition.

There are many kinds of change that are merely
another name for simple remission of the intellectual
strain. When a severe and difficult exercise is exchanged
for an easy one, the agreeable effect is due, not to what
we engage in, but to what we are relieved from. For
letting down the strain of the faculties, it is sometimes
better to take up a light occupation for a time than to
be totally idle.

The exchange of study for sport has the twofold
advantage of muscular exercise and agreeable play. To
pass from anything that is simply laborious to the
indulgence of a taste or liking, is the fruition of life.
To emerge from constraint to liberty, from the dark to
the light, from monotony to variety, from giving to receiving—is the exchanging of pain for pleasure. This,
which is the substantial reward of labour, is also the
condition of renovating the powers for further labour
and endurance.

To come closer to the difficulty in hand. The kind
of change that may take place within the field of study
itself, and that may operate both as a relief from strain

and as the reclamation of waste ground, is best exemplified in such matters as these :—In the act of learning
generally there is a twofold attitude—observing what is
to be done, and doing it. In verbal exercises we first
listen and then repeat ; in handicraft, we look at the
model, and then reproduce it. Now, the proportioning
of the two attitudes is a matter of economical adjustment. If we are kept too long on the observing stretch,
we lose the energy for acting ; not to mention that more
has been given us than we are able to realize. On the
other hand, we should observe long enough to be quite
saturated with the impression ; we should have enough
given us to be worthy of our reproducing energy. Anyone working from a model at command learns the suitable proportion between observing and doing. The living
teacher may err on either side. He may give too much
at one dose ; this is the common error. Or he may
dole out insignificantly small portions, such as do not
evoke the sense of power in the pupils.

When an arduous combination is once struck out,
the worst is over, but the acquisition is not completed.
There is the farther stage of repetition and practice, to
give facility and to ensure permanence. This is comparatively easy. It is the occupation of the soldier after his
first year. There is a plastic process still going on, but
it is not the same draft upon the forces as the original
struggles. At this stage, other acquirements are possible
and should be made. Now, in the course of training, it
is a relief to pass from the exercises that are entirely
new and strange to those that have been practised and
need only to be continued and confirmed.

Before considering the alternations of departments

of acquisition, we may advert to the two different intellectual energies, called, respectively, Memory and Judgment. These are in every way distinct, and in passing
from the one to the other, there is a real, and not merely
an apparent, transition. Memory is nearly identical with
the Retentive, Adhesive, or Plastic faculty, which I have
assumed to be perhaps the most costly employment of
the powers of the mind and brain. Judgment again
may be simply an exercise of Discrimination; it may
also involve Similarity and Identification; it may further
contain a Constructive operation. It is the aspect of
our intellectual power that turns to account our existing
impressions, as contrasted with the power that adds to
our accumulated stores. The most delightful and fructifying of all the intellectual energies is the power of
Similarity and Agreement, by which we rise from the
individual to the general, trace sameness in diversity,
and master, instead of being mastered by, the multiplicity of Nature.

Much more would be necessary to exhaust the nature
of the opposition between exercises of Memory and exercises of Judgment. Language and Science approxi-
mately represent the contrast, although language does
not exclude judgment; and science demands memory.
But in the one region, mere adhesion is in the ascendant,
and, in the other, the detection of similarity in diversity
is the leading circumstance. There is thus a real transition, and change of strain, in passing from the one
class of studies to the other; the only qualifying circumstance is, that in early years routine adhesion plays
the greatest part, being, in fact, easier than the other line
of exertion, for reasons that can be divined.

We can now see what are the departments that con·
stitute the most effective transitions or diversions whereby relief may be gained at one point, and acquirement
pushed at some other. In the muscular acquirements,
we have several distinct regions—the body generally,
the hand in particular, the voice (articulate) and the
voice (musical). To pass from one of these to the other
is almost a total change. Then, as to the Sense engaged,
we may alternate between the eye and the ear, making
another complete transition. Further, each of the sense-
organs has distinguishable susceptibilities ; as colour and
form to the eye, articulation and music to the ear.

Another effective transition is from books or spoken
teaching to concrete objects, as set forth in the sciences
of observation and experiment. The change is nearly
the same as from an abstract subject, like Mathematics,
to one of the concrete and experimental sciences, as
Botany or Chemistry. A still further change is from
the world of matter to the world of mind ; but this is
liable to assume false and delusive appearances.

It has been well remarked that Arithmetic is an
effective transition from Reading and Writing. The
whole strain and attitude of the mind is entirely different, when the pupil sets to perform sums after a reading lesson. The Mathematical sciences are naturally
deemed the driest and hardest of occupations to the
average mind ; yet there may be occupations such as
to make them an acceptable diversion. I have known
clergymen whose relaxation from clerical duty consisted
in algebraical and geometrical problems.

The Fine Art acquisitions introduce an agreeable
variety, partly by bringing distinctive organs into play,

and partly by evolving a pleasurable interest that enters little, if at all, into other studies. The more genial part of Moral Training has a relationship to Art ; the severer exercises are a painful necessity, and not an agreeable transition from anything.

The introduction of narratives, stirring incidents, and topics of human interest generally, is chiefly a mode of pleasurable recreation. If taken in any other view, it falls under some of the leading studies, and engages the Memory, the Judgment, or the Constructive power, and must be estimated accordingly.

Bodily training, Fine Art (itself an aggregate of alternations), Language, Science, do not exhaust all the varieties of acquirement, but they indicate the chief departments whose alternation gives relief to the mental strain, and economizes power in the whole. Under these, as already hinted, there are variations of attitude and exercise ; from listening to repeating, from learning a rule to the application of it in new cases, from knowledge generally to practice.

The transition from one language to another, being a variation in the nature of the impressions, is a relief of an inferior kind, yet real. It is the more so, if we are not engaged in parallel exercises ; learning strings of Latin words in the morning, and of German in the evening, does not constitute any relief.

From one science to another the transition may be great, as already shown, or it may be small. From Botany to Zoology affords a transition of material, with similarity in form. Pure and Mixed Mathematics are the very same thing. The change from Algebra to Geometry is but slightly refreshing ; from Geometry to

Trigonometry and Geometrical Conic Sections, is no relief to any faculty.

There are minor incidents of relief and alternation that are not to be despised. Passing from one master to another (both being supposed competent) is a very sensible and grateful change ; even the change of room, of seat, of posture, is an antidote against weariness, and helps us in making a fresh start. The jaded student relishes a change of books in the same subject.[1]

Some subjects are in themselves so mixed that they would appear to contain the elements of a sufficiently varied occupation of the mind ; such are Geography, History, and what is called Literature, when studied both for expression and for subject-matter. This variety, however, is not altogether a desirable thing. The analytic branch of the Science of Education would have to resolve those aggregates into their constituent parts, and to consider not only their respective contributions to our mental culture, but also the advantages and disadvantages attending the mixture.

CULTURE OF THE EMOTIONS.

The laws attainable in the departments of Emotion and Volition are the immediate prelude to Moral Education, in which all the highest difficulties culminate. There are emotional and volitional forces prior to any cultivation, and there are new forces that arise through cultivation ; yet from the vagueness attaching to the

[1] In the German Gymnasia, where the routine is very strict, and the exactions enormous, the pupils are allowed a day in the week to their own choice of studies.

measured intensity of feelings and emotions, it is not easy to value the separate results.

The general laws of Retentiveness equally apply to emotional growths. There must be Repetition and Concentration of mind to bring about a mental association of pleasure or of pain with any object. But there are peculiarities in the case such as to demand for it a supplementary treatment. Perhaps the best way of bringing out the points is to indicate the modes or species of growths, coming under Emotion and Volition, that most obtrude themselves upon the notice of the educationist.

1. We may quote first the Associations of Pleasure and Pain with the various things that have been present to us during our experiences of delight and suffering. It is well known that we contract pleasurable regards towards things originally indifferent that have been often present to us in happy moments. Local associations are among the most familiar examples; if our life is joyous, we go on increasing our attachments to our permanent home and neighbourhood ; we are severely tried when we have to migrate ; and one of our holiday delights is to revisit the scenes of former pleasures. A second class of acquired feelings includes the associations with such objects as have been the instruments of our avocations, tastes, and pursuits. The furnishings of our home, our tools, weapons, curiosities, collections, books, pictures—all contract a glow of associated feeling, that helps to palliate the dulness of life. The essence of affection, as distinguished from emotion, is understood to be the confirming and strengthening of some primary object of our regards. As our knowledge extends, we contract numerous associations with things purely ideal,

as with historic places, persons, incidents. I need
only allude to the large field of ceremonies, rites, and
formalities that are cherished as enlarging the surface
of emotional growths. The Fine Art problem of distinguishing between original and derived effects consists in
more precisely estimating these acquired pleasures.

The educationist could not but cast a longing eye
over the wide region here opened up, as a grand opportunity for his art. It is the realm of vague possibility,
peculiarly suited to sanguine estimates. An education
in happiness pure and simple, by well-placed joyous associations, is a dazzling prospect. One of Sydney Smith's
pithy sayings was —'If you make children happy now,
you make them happy twenty years hence, by the memory of it.' This referred no doubt to the home life. It
may, however, be carried out also in the school life ; and
enthusiasm has gone the length of supposing that the
school may be so well constituted as to efface the stamp
of an unhappy home.

The growth of such happy associations is not the
work of days; it demands years. I have endeavoured
to set forth the psychology of the case,[1] and do not here
repeat the principles and conditions that seem to be involved. But the thread of the present exposition would
be snapt, if I were not to ask attention to the difference
in the rate of growth when the feelings are painful ; in
which case, the progress is not so tedious, nor is it so
liable to thwarting and interruption.

With understood exceptions, pleasure is related phy-
sically with vitality, health, vigour, harmonious adjustment of all the parts of the system ; it needs sufficiency

---

[1] *The Emotions and the Will*, 3rd edit., p. 89.

of nutriment or support, excitement within due limits, the absence of everything that could mar or irritate any organ. Pain comes of the deficiency in any of these conditions, and is, therefore, as easy to bring about and maintain as the other is difficult. To evoke an echo or recollection of pleasure, is to secure, or at least to simulate, the copiousness, the due adjustment and harmony, of the powers. This may be easy enough when such is the actual state at the time, but that is no test. What we need is to induce a pleasurable tone, when the actuality is no more than indifferent or neutral, and even, in the midst of actual pain, to restore pleasure by force of mental adhesiveness. A growth of this description is, on *à priori* grounds, not likely to be soon reached.

On the other hand, pain is easy in the actual, and easy in the ideal. It is easy to burn one's fingers, and easy to associate pain with a flame, a cinder, a hot iron. Going as spectators to visit a fine mansion, we feel in some degree elated by the associations of enjoyment ; but we are apt to be in a still greater degree depressed by entering the abodes of wretchedness, or visiting the gloomy chambers of a prison.

2. The facility of painful growths is not fully comprehended, until we advert to the case of Passionate Outbursts, or the modes of feeling whose characteristic is Explosiveness. These costly discharges of vital energy are easy to induce at first hand, and easy to attach to indifferent things, so as to be induced at second hand likewise. Very rarely are they desirable in themselves ; our study is to check and control them in their original operation, and to hinder the rise of new occasions for their display One of the best examples is Terror ; an

explosive and wasteful manifestation of energy under certain forms of pain. If it is frequently stimulated by its proper causes, it attaches itself to bystanding circum-stances with fatal readiness, and proceeds with no tardy steps. Next is Irascibility; also an explosive emotion. It too, if ready to burst out by its primary causes, soon enlarges its borders by new associations. It is in every way more dangerous than terror. The state of fear is so miserable that we would restrain it if we could. The state of anger, although containing painful elements, is in its nature a luxurious mood ; and we may not wish either to check it in the first instance, or to prevent its spreading over collateral things. When anyone has stirrred our irascibility to its depths, the feeling overflows upon all that relates to him. If this be plea-sure, it is a pleasure of rapid growth ; even in tender years we may be advanced in hatreds. That combina-tion of terror and irascibility giving rise to what is named Antipathy is (unless strongly resisted) a state easy to assume and easy to cultivate, and is in wide contrast with the slow growth of the pleasures typified under the foregoing head. A signal illustration of explosiveness is furnished by Laughter, which has both its original causes, and its factitious or borrowed stimulants. This is an instance where the severity of the agitation pro-vokes self-control, and where advancing years contract rather than enlarge the sphere. As the expression of disparaging and scornful emotions, its cultivation has the facility of the generic passion of malevolence. We may refer, next, to the explosive emotion of Grief, which is in itself seductive, and, if uncontrolled, adds to its primary urgency the force of a habit all too readily

6

acquired. There is, moreover, in connection with the
Tender Emotion, an explosive mode of genuine affection, of which the only defect is its being too strong to
last ; it prompts to a degree of momentary ardour that
is compatible with a relapse into coldness and neglect.
This, too, will spontaneously extend itself, and will exemplify the growth of emotional association with undesirable rapidity.

What has now been said is but a summary and representation of familiar emotional facts. Familiar also
is the remark that explosiveness is the weakness of early
life, and is surmounted to a great degree by the lapse of
time and the strengthening of the energies. The encounter with others in every-day life begets restraint
and control ; and one's own prudential reflections stimulate a further repression of the original outbursts, by
which also their growth into habits is retarded. In so
far as they are repressed by influence from without,
and counter-habits established, as a part of moral education, I have elsewhere stated what I consider the
two main conditions of such a result—a powerful initia-
tive, and an unbroken series of conquests. When
these conditions are exemplified through all the emotions in detail, the specialities of the different genera
—Fear, Anger, Love, and the rest—are sufficiently
obvious.

3. The chief interest always centres in those associations that, from their bearing on conduct as right and
wrong, receive the name 'Moral.' The class just described have this bearing in a very direct form ; while
the bearing of the first class is only indirect. But when
we approach the subject with an express view to moral

culture, we must cross the field of emotional association in general by a new track.

Moral improvement is obviously a strengthening of this so-called Moral Faculty, or Conscience—increasing its might (in Butler's phrase) to the level of its right. But in order to strengthen an energy we must know what it is: if it is a simple, we must define it in its simplicity ; if it is a compound, we must assign its elements, with a view to defining them. The unconventional handling of moral culture by Bentham and by James Mill is strongly illustrative of this part of the case. Mill's view of the Moral Sense is the theory of thorough-going derivation ; and, in delineating the process of Moral Education, he naturally follows out that view. He takes the cardinal virtues piecemeal ; for example :—' Temperance bears a reference to pain and pleasure. The object is, to connect with each pain and pleasure those trains of ideas which, according to the order established among events, tend most effectually to increase the sum of pleasures upon the whole, and diminish that of pains.' The advocates of a Moral Faculty would have a different way of inculcating Temperance, which, however, I will not undertake to reproduce.

It will not be denied, as a matter of fact, that the perennial mode of ensuring the moral conduct of mankind has been punishment and reward—pain and pleasure. This method has been found, generally speaking, to answer the purpose ; it has reached the springs of action of human beings of every hue. No special mental endowment has been needed to make man dread the pains of the civil authority. Constituted as we are to flee all sorts of pain, we are necessarily urged to avoid pain

when it comes as punishment. Education is not essential to this effect, any more than it is essential to our
avoiding the pains of hunger, cold, or fatigue.

Those that demur to the existence of a special faculty,
different from all the other recognized constituents of
mind—Feeling, Will, or Intellect—are not to be held
as declaring that Conscience is entirely a matter of education ; for, without any education at all, man may be, to
all intents and purposes, moral. What is meant by the
derivative theory of Conscience is, that everything that
it includes is traceable to some one or other of the leading facts of our nature ; first of all to Will or Volition,
motived by pain and pleasure, and next to the Social
and Sympathetic impulses. The co-operation of these
factors supplies an impetus to right conduct that is nearly
all-powerful, wherever there is the external machinery of
law and authority. Education, as a third factor, plays a
part, no doubt, but we may over-rate as well as under-
rate its influence. I should not be far out in saying
that seventy-five per cent. of the average moral faculty
is the rough and ready response of the Will to the constituted penalties and rewards of society.

At the risk of embroiling the theory of Education in
a controversy that would seem to be alien to it, I conceive it to be necessary to make these broad statements
as a prelude to inquiring what are the emotional and
volitional associations that constitute the made-up or
acquired portion of our moral nature. That education is
a considerable factor is shown by the difference between
the children that are neglected and those that are carefully tended ; a difference, however, that means a good
deal more than education.

When the terrors of the law are once thoroughly understood, it does not seem as if any education could add
to the mind's own original repugnance to incur them;
and, on the other hand, when something in the nature of
reward is held forth to encourage certain kinds of conduct, we do not need special instruction to prompt us to
secure it.   There is, indeed, one obvious weakness that
often nullifies the operation of these motives, namely, the
giving way to some present and pressing solicitation ; a
weakness that education might do something for, but
rarely does.   The instructor that could reform a victim
to this frailty, would effect something much wider than
is properly included in moral improvement.

Going in search of some distinct lines of emotional
association that enhance the original impulses coincident
with moral duty, I think I may cite the growth of an
immediate, independent, and disinterested repugnance
to what is uniformly denounced and punished as being
wrong.   This is a state or disposition of mind forming
part of a well-developed conscience ; it may grow up
spontaneously under the experience of social authority,
and it may be aided by inculcation ; it may, however,
also fail to show itself.   It is the parallel of the much-
quoted love of money for itself; but is not so facile in
its growth.   For one thing, the mind must not treat
authority as an enemy to be counted with, and to be
obeyed only when we cannot do better.   There must
be a cordial acquiescence in the social system as
working by penalties; and this needs the concurrence
of good impulses with reflection on the evils that mankind are rescued from.   It is by being favourably situated in the world, as well as by being sympathetically

disposed, that we contract this repugnance to immoral
acts in themselves, without reference to the penalties
that are behind ; and thus perform our duties when out
of sight, not in the narrowness of the letter, but in the
fulness of the spirit. It would take some consideration to show how the schoolmaster might co-operate in
furthering this special growth.

In Education there has to be encountered at every
turn the play of Motives. Now, the theory of Motives
is the theory of Sensation, Emotion and Will ; in other
words, it is the psychology of the Sensitive and the
Active Powers.

PLAY OF MOTIVES :—THE SENSES.

The pleasures, the pains and the privations of the
Senses are the earliest and the most unfailing, if not also
the strongest, of motives. Besides their bearings on
self-preservation, they are a principal standing dish in
life's feast.

It is when the Senses are looked at on the side of
feeling, or as pleasure and pain, that the defectiveness of
the current classification into five is most evident. For,
although, in the point of view of knowledge or intellect,
the five senses are the really important approaches to
the mind, yet, in the view of feeling or pleasure and pain,
the omission of the varied organic susceptibility leaves
a wide gap in the handling of the subject. Some of
our very strongest pleasures and pains grow out of
the region of organic life—Digestion, Circulation, Respiration, Muscular and Nervous integrity or derange·
ment.

In exerting influence over human beings this department of sensibility is a first resource. It can be counted
on with more certainty than perhaps any other. Indeed,
almost all the punishments of a purely physical kind fall
within the domain of the organic sensations. What is
it that makes punishment formidable but its threatening
the very vitals of the system? It is the lower degree of
what, in a higher degree, takes away life.

For example, the Muscular System is the seat of a
mass of sensibility, pleasurable and painful: the pleasures of healthy exercise, the pains of privation of exercise, and the pains of extreme fatigue. In early life,
when all the muscles, as well as the senses, are fresh,
the muscular organs are very largely connected both
with enjoyment and with suffering. To accord full scope
to the activity of the fresh organs is a gratification that
may take the form of a rich reward; to refuse this scope
is the infliction of misery; to compel exercise beyond
the limits of the powers is still greater misery. Our
penal discipline adopts the two forms of pain: in the
milder treatment of the young, the irksomeness of restraint; in the severer methods with the full-grown, the
torture of fatigue.

Again, the Nervous System is subject to organic depression; and certain of our pains are due to this cause.
The well-known state denominated 'Tedium' is nervous
uneasiness; and is caused by undue exercise of any
portion of the nervous system. In its extreme forms, it
is intolerable wretchedness. It is the suffering caused
by penal impositions or tasks, by confinement, and by
monotony of all kinds. The acute sufferings of the
nervous system, as growing out of natural causes, are

represented by neuralgic pains.   It is in graduated
artificial inflictions, operating directly on the nerves by
means of electricity, that we may look for the physical
punishments of the future, that are to displace floggings
and muscular torture.

The interests of Nourishment, as against privation
of food, are necessarily bound up with a large volume
of enjoyment and suffering.   Starvation, deficiency and
inferiority of food, are connected with depression and
misery of the severest kind, inspiring the dread that most
effectually stimulates human beings to work, to beg, or
to steal.   The obverse condition of a rich and abundant
diet is in itself an almost sufficient basis of enjoyment.
The play of motives between those extremes enables us
to put forth an extensive sway over human conduct.

An instructive distinction may be made between
Privation and Hunger ; likewise between their opposites.
Privation is the positive deficiency of nourishing material
in the blood ; Hunger is the craving of the stomach at
its usual times of being supplied, and is a local sensibility, perhaps very acute, but not marked by the profound wretchedness of inanition.   There may be plenty
of material to go on with, although we are suffering from
stomachic hunger.   Punishing, for once, by the loss of a
meal out of the three or four in the day is unimportant
as regards the general vigour, yet very telling as a
motive.   Absolutely to diminish the available nutriment
of the system is a measure of great severity ; to inflict a
passing hunger is not the same thing.

When we unite the acute pleasures of the palate with
stomachic relish and the exhilaration of abundance of
food-material in a healthy frame, we count up a large

mass of pleasurable sensibility. Between the lowest demands of subsistence and the highest luxuries of affluent means there is a great range, available as an instrumentality of control in the discipline of the young. The usual regimen being something considerably above necessaries, and yet beneath the highest pitch of indulgence, room is given to operate both by reduction and by increase of luxury, without either mischief or pampering; and as the sensibility in early years is very keen under those heads, the motive power is great. Having in view the necessities of discipline with the young, the habitual regimen in food should not be pitched either too low or too high to permit of such variations. It is the misfortune of poverty that this means of influence is greatly wanting; the next lower depth to the delinquent child is the application of the rod.

These are the chief departments of Organic Sensibility that contain the motives made use of in reward and punishment. The inflictions of caning and flogging operate upon the organ of the sense of touch, yet, in reality, the effect is one to be classed among the pains of organic life, rather than among tactile sensations; it is a pain resulting from injury or violence to the tissue in the first instance, and, if carried far, is destructive of life. Like all physical acute pains, it is a powerful deterring influence, and is doubtless the favourite punishment of every age and every race of mankind. The limitations to its use demand a rigorous handling; but the consideration of these is mixed up with motives afterwards to be adverted to.

The ordinary five Senses contain, in addition to their intellectual functions, many considerable sensibilities to

pleasure and pain.   The pleasures can be largely made
use of as incentives to conduct.   The pains might, of
course, be also employed in the same way ; but, with the
exceptions already indicated, they very rarely are.   We
do not punish by bad odours, nor by bitter tastes.
Harsh and grating sounds may be very torturing, but
they are not used in discipline.   The pains of sight reach
the highest acuteness, but as punishment they are found
only in the most barbarous codes.

Postponing a review of the principles of punishment
generally, we approach the most perplexing department
of motives—the higher Emotions.   Few of the simple
sensational effects are obtained in purity, that is, without
the intermingling of emotions.

PLAY OF MOTIVES :—THE EMOTIONS.

One large department of Psychology is made up of
the classification, definition, and analysis of the Emotions.
The applications of a complete theory of Emotion are
numerous, and the systematic expansion must be such
as to cope with all these applications.   We here narrow
the subject to what is indispensable for the play of
motives in Education.

First of all, it is necessary to take note of the large
region of Sociability, comprising the Social Emotions
and Affections.   Next is the department of anti-social
feeling—Anger, Malevolence, and Lust of Domination.
Taking both the sources and the ramifications of these
two leading groups, we cover perhaps three-fourths of all
the sensibility that rises above the Senses proper.   They
do not, indeed, exhaust the fountains of emotion, but they

leave no others that can rank as of first-class importance, except through derivation from them and the senses together.

The region of Fine Art comprises a large compass of pleasurable feeling, with corresponding susceptibilities to pain. Some of this is sensation proper, being the pleasures of the two higher senses; some of it is due to associations with the interests of all the senses (Beauty of Utility); a certain portion may be called intellectual — the perception of unity in variety; whilst the still largest share appears to be derived from the two great sources above described.

The Intellect generally is a source of various gratifications and also of sufferings that are necessarily mixed up with our intellectual education. Both the delights of attained knowledge and the pains of intellectual labour have to be carefully counted with by every instructor.

The pleasures of Action or Activity are a class greatly pressed into the educational service, and, therefore, demand special consideration.

The names Self-esteem, Pride, Vanity, Love of Praise, express powerful sentiments, whose analysis is attended with much subtlety. They are largely appealed to by everyone that has to exercise control over human beings. To gratify them, is to impart copious pleasure; to thwart or wound them, is to inflict corresponding pain.

Mention has not yet been made of one genus of emotion, formidable as a source of pain and as a motive to activity, namely, Fear or Terror. Only in the shape of reaction or relief, is it a source of pleasure. The skilful management of this sensibility has much to do with the

efficient control of all sentient creatures, and still more
with the saving of gratuitous misery.

Our rapid review of these various sources of emotion,
together with others of a minor kind, proposes to deal
once for all, and in the best manner, with the various
educational questions that turn upon the operation of
motives. We shall have to remark upon prevailing exaggerations on some heads, and upon the insufficient
stress laid on others ; and we shall endeavour to unfold
in just proportions the entire compass of our emotional
susceptibilities available for the purposes of the teacher.

The Emotion of Terror.

The state of mind named Terror or Fear is described
briefly as a state of extreme misery and depression,
prostrating the activity and causing exaggeration of
ideas in whatever is related to it. It is an addition to
pain pure and simple—the pain of a present infliction.
It is roused by the foretaste or prospect of evil, especially
if that is great in amount, and still more if it is of uncertain nature.

As far as Education is concerned, terror is an incident
of the infliction of punishment. We may work by the
motive of evil without producing the state of terror, as
when the evil is slight and well-defined ; a small understood privation, a moderate dose of irksomeness, may be
salutary and preventive, without any admixture of the
quakings and misery of fear. A severe infliction in
prospect will induce fear ; the more so that the subject
does not know how severe it is to be.

In the higher moral Education, the management of

Fear is of the utmost consequence. So great are the evils attendant on the use of it, that it should be reserved for the last resort. Fear wastes the energy and scatters the thoughts, and thus is ruinous to the interests of mental progress. Its one certain result is to paralyze and arrest action, or else to concentrate force in some single point, at the cost of general debility. The tyrant, working by terror, disarms rebelliousness, but fails to procure energetic service.

The worst of all modes and instruments of discipline is the employment of spiritual, ghostly, or superstitious terrors. Unless it were to scourge and thwart the greatest of criminals—the disturbers of the peace of mankind—hardly anything justifies the terrors of superstition. On a small scale, we know what it is to frighten children with ghosts ; on a larger scale, we have the influence of religions dealing almost exclusively in the fear of another life.

Like the other gross passions, Terror admits of being renned upon and toned down, till it becomes simply a gentle stimulation ; and the reaction more than compensates for the misery. The greatest efforts in this direction are found in the artistic handling of fear, as in the sympathetic fears of tragedy, and in the passing terrors of a well-constructed plot. In the moral bearings of the emotions, its refined modes are shown in the fear of giving pain or offence to one that we love, respect, or venerate. There may be a considerable degree of the depressing element even in this situation ; yet the effect is altogether wholesome and ennobling. All superiors should aspire to be feared in this manner.

Timiditv, or susceptibility to fear, is one of the noted

differences of character ; and this difference is to be
taken into account in discipline. The absence of general vigour, bodily and mental, is marked by timidity ;
and the state may also be the result of long bad usage,
and of perverted views of the world. In the way of
culture, or of high exertion in any form, little is to be
expected from thoroughly timid natures ; they can be
easily governed, so far as concerns sins of commission,
but their omissions are not equally remediable.

The conquest of superstitious fears is one of the
grandest objects of education taken in its widest compass. It cannot be accomplished by any direct inculcation ; it is one of the incidental and most beneficial
results of the exact study of nature, in other words,
science.

The Social Motives.

This is perhaps the most extensive and the least
involved of all the emotional influences at work in
Education.

The pleasures of Love, Affection, Mutual Regard,
Sympathy, or Sociability, make up the foremost satisfaction of human life ; and, as such, are a standing object
of desire, pursuit, and fruition. Sociability is a wholly
distinct fact from the prime supports of existence and
from the pleasures of the five senses, and is not, in my
opinion, resolvable into those, however deeply we may
analyze it, or however far back we may trace the historical evolution of the mind. Nevertheless, as the supports
of life and the pure sense agreeables and exemptions,
come to us in great part through the medium of fellow-
beings, the value of the social regards receives from this

cause an enormous augmentation, and, in the total, counts for one paramount object of human solicitude. It would appear strange if this motive could ever be overlooked by the educator, or by anyone; yet there are theories and methods that treat it as of inferior account.

The vast aggregate of social feeling is made up of the intenser elements of sexual and parental love, and the select attachments in the way of friendship, together with the more diffused sentiments towards the masses of human beings. The motive power of the feelings in education may be well exemplified in the intense examples; we can see in these both the merits and the defects of the social stimulus. The *Phædrus* of Plato is a remarkable ideal picture of philosophy prompted by Eros, in the Grecian form of attachment. The ordinary love of the sexes, in our time, does not furnish many instances of the mutual striving after high culture; it may be left out of account in the theory of early education. We frequently find mothers applying to studies that they feel no personal attraction for, in order to assist in the progress of their children. This is much better than nothing; a secondary end may be the initiation and discovery of a taste that at last is self-subsisting.

The intense emotions, from the very fact of their intensity, are unsuited to the promptings of severe culture. The hardest studious work, the laying of foundations, should be over before the flame of sexual and parental passion is kindled; when this is at its height the intellectual power is in abeyance, or else it is diverted from its regular course. The mutual influence of two lovers is not educative, for want of the proper conditions

No doubt considerable efforts are inspired ; but there is seldom sufficient elevation of view on the one side, or sufficient adaptability on the other, to make the mutual influence what Plato and the romancists conceive as possible. By very different and inferior compliances on both sides, the feeling may be kept alive ; if more is wanted, it dies away.

The favourable conjunction for study and mental culture in general is friendship between two, or a small number, each naturally smitten with the love of know-ledge for its own sake, and basing their attachment on that circumstance. A certain amount of mutual liking in other respects perfects the relationship ; but the over-powering sensuous regards of the Platonic couple do not furnish the requisite soil for high culture. As a matter of fact, those attachments, as they existed in Greece, prompted to signal instances of self-devotion in the form of surrendering worldly goods and life itself ; and this is the highest fruit that they have yielded in later times.

The remaining aspect of sociability—the influence of the general multitude—holds out the most powerful and permanent motive to conduct, and is largely felt in edu-cation. In the presence of an assembly the individual is roused, agitated, swayed ; the thrill of numbers is electric ; in whatever direction the influence tends, it is almost irresistible. Any effort made in the sight of a host is, by that circumstance, totally altered in charac-ter ; and all impressions are very much deepened.

Having in view this ascendancy of numbers, we can make a step towards computing the efficacy of class teaching, public schools, and institutions where great multitudes are brought together. The power exercised

is of a mixed character; and the several elements admit
of being singled out. The social motive, in its pure
form of gregarious attraction and mutual sympathy,
does not stand alone. Supposing it did, the effect would
be to supply a strong stimulus in favour of everything
that was supported by common consent; the individual
would be urged to attain the level of the mass. The
drill of a regiment of soldiers corresponds very nearly
to this situation; every man is under the eye of the
whole, and aspires to be what the rest are, and not
much, if anything, beyond: the sympathetic co-operation of the mass guides, stimulates, and rewards the
exertion of the individual. Even if it were the distina-
tion of a soldier to act as an isolated individual, still
his education would be most efficaciously conducted on
the mass system; being finished off by a certain amount
of separate exercise to prepare for the detached or independent position.

In all cases of teaching numbers together, the social
feeling, in the pure form now assumed, is frequently
operative; and the results are as stated. The tendency
is to secure a certain approved level of attainment; those
that are disinclined of themselves to work up to that
level are pushed on by the influence of the mass. If
there were no other strong passions called out in society,
the general result would be a kind of Communism or
Socialism characterized by mediocrity and dead level;
everything correct up to a certain point, but no individual superiority or distinction.

The influence of society as the dispenser of collective
good and evil things, in addition to its operation in the
affections and the sympathies, is necessarily all-powerful in

7

every direction.   If this stimulus were always to coincide
with high mental culture, the effect would be something
that the imagination hardly dares to shadow forth.   It
is, however, a power that may be propitiated by many
different means, including shams and evasions ; and the
bearing upon culture is only occasional.   Nevertheless,
the social rewards have often served to foster the highest
genius— the oratory of Demosthenes, and the poetry of
Horace and of Virgil—that form of genius that is noto-
riously allied with toil and perseverance of the most
arduous kind.   The same influence, working by disapprobation and approbation combined, is, as I contend,
the principal generating source of the ordinary moral
sentiments of mankind, and the inspiration of exceptional virtues.

The Anti-Social and Malign Emotions.

The emotions of Anger, Hatred, Antipathy, Rivalry,
Contumely, have reference to other beings, no less than
Love or Affection ; but in an opposite way.   In spite of
the painful incidents in their manifestation—the offence in
the first instance, and next the dangers of reprisal—they
are a source of immediate pleasure, often not inferior,
and sometimes superior, in amount to the pleasures of
amity and gregarious co-operation.   In numerous instances people are willing to forego social and sympathetic delights to indulge in the pleasures of malignity.

In the work of discipline the present class of emotions
occasion much solicitude.   They can in certain ways
be turned to good account, but for the most part the

business of the educator and the moralist is to counter-work them, as being fraught with unalloyed evil.

Being a fitful or explosive passion, Anger should, as far as possible, be checked or controlled in the young; but there are no adequate means of doing so short of the very highest influence of the parent or the teacher. The restraint induced by the presence of a dread superior at the time does not sink deep enough to make a habit; opportunities are sought and found to vent the passion with safety. The cultivation of the sympathies and the affections is what alone copes with angry passion, whether we take it as a disturber of equanimity, or as the prompter of wrong. The obverse of ill-temper is the disposition that thinks less of harm done to self, and more of harm done to other people; and if we can do anything to foster this disposition, we thereby reduce the sphere of malignant passion. The collateral incentives to suppress angry passion include, besides the universal remedy of disapprobation, an appeal to the sense of personal dignity and to the baneful consequences of passionate outbursts. ·

The worst form of malignant feeling is cold and deliberate delight in cruelty; all too frequent, especially in the young. The torturing of animals, of weak and defenceless human beings, is the spontaneous outflow of the perennial fountain of malevolence. This has to be checked, if need be, at the expense of considerable severity. The inflictions practised on those that are able to recriminate generally find their own remedy; and the discipline of consequences is as effectual as any. By having to fight our equals, we are taught to regulate our wrathful and cruel propensities.

The intense pleasure of victory contains the sweet-ness of malevolence, heightened by some other ingre-dients. The prostration and destruction of an enemy or a rival is, no doubt, the primary situation where male-volent impulses had their rise; and it continues to be perhaps the very strongest stimulant of the human energies. Notwithstanding its several drawbacks, we are obliged to give it a place among motives to study and mental advancement. In the fight and struggle of party contests, the pleasure of victory enters in full flavour; and in the competitions at school, the same motive is at work. The social problem of restraining individuals in their selfish grasping at good things—the mere agreeables and exemptions of the senses—is ren-dered still more intractable by the craving for the smack of malevolent gratification. Total repression has been found impossible; and ingenuity has devised a number of outlets that are more or less compatible with the sacredness of mutual rights.

One chief outlet for the malevolent impulses is the avenging of wrong, whether private or public. A con-victed wrongdoer is punished by the law, and the indig-nation roused by the crime turns to gratification at the punishment. In the theory of penal retribution some allowance is claimed for the vindictive satisfaction of the public. To think only of the prevention of crime and the reformation of criminals, suppressing all re-sentful feeling, is a thing too severe and ascetic for human nature as at present constituted. The privacy of the punishments of criminals, in our modern system, is intended to keep the indulgence within bounds.

A wide ideal scope is given to our resentful pleasures

in history and in romance ; we are gratified by the retribution inflicted upon the authors of wrong.   Narratives of evildoers and of their punishment are level to
the meanest capacity ; this is the sort of history that
suits even the imagination of children.

The highest refinement of the malevolent gratification I take to be the emotion called the Ludicrous and
the Comic.   There is a laugh of vindictiveness, hatred,
and derision, that carries the sentiment as far as it can
be carried without blows.   But there is also the laugh
expressed by Playfulness and Humour, in which the
malignant feeling seems almost on the point of disappearing in favour of the amicable sentiment.   It is of
some importance to understand that in play, fun, and
humour there is a delicate counterpoise of opposing
sentiments, an attempt to make the most of both worlds
— Love and Anger.   The great masterpieces of humour
in literature, the amenities of everyday society, the innocent joyousness of laughter—all attest the success of
the hazardous combination.   Nothing could better show
the intensity of the primitive charm of malevolence
than the unction that survives after it is attenuated to
the condition of innocent mirthfulness.   When the real
exercise of the destructive propensity is not to be had,
creatures endowed with emotions still relish the fictitious
forms.   This is seen remarkably in the amicable 'play'
of puppies and kittens.   Not being endowed with much
compass of the caressing acts, they show their love by
snarling and sham biting ; in which, through their for
tunate self-restraint, they seem to enjoy a double pleasure.
In the play of children there is the same employment
of the forms of destructive malevolence, and, so long as

it is happily balanced, the effect is highly piquant. By submitting in turn to be victimized, a party of children can secure, at a moderate cost to each, the zest of the malevolent feeling ; and this I take to be the quintessence of play.

The use of this close analysis is to fix attention upon the precarious tenure of all these enjoyments, and to render a precise reason for the well-known fact that play or fun is always on the eve of becoming earnest ; in other words, the destructive or malevolent element is in constant danger of breaking loose from its checks, and of passing from fictitious to actual inflictions. The play of the canine and the feline kind often degenerates in this fashion ; and in childish and youthful amusements it is a perpetual rock ahead.

It is no less dangerous to indulge people in too much ideal gratification of the vindictive sentiments. Tales of revenge against enemies are too apt to cultivate the malevolent propensity. Children, it is true, take up this theme with wonderful alacrity ; nevertheless, it is a species of pampering supplied to the worst emotions instead of the best.

One other bearing of Irascibility on Education needs to be touched upon. When disapproval is heightened with Anger, the dread inspired is much greater. The victim anticipates a more severe infliction when the angry passion has been roused ; hence the supposition is natural that anger is an aid to discipline. This, however, needs qualifying. Of course any increase of severity has a known deterrent effect, with whatever drawbacks may attend the excess. But anger is fitful ; and, therefore, its co-operation mars discipline by want of measure, and want of consistency; when the fit has passed, the mind often relapses into a mood unfavourable to a proper amount of repression.

The function of anger in discipline may be something very grand, provided the passion can be controlled. There is a fine attitude of indignation against wrong that may be assumed with the best effect. It supposes the most perfect self-command, and is no more excited than seems befitting the occasion. Mankind would not be contented to see the bench of Justice occupied by a calculating machine that turned up a penalty of five pounds, or a month's imprisonment, when certain facts were dropped in at the hopper. A regulated expression of angry feeling is a force in itself. Neither containing fitfulness, nor conducting to excess of infliction, it is the awe-inspiring personation of Justice, and is often sufficient to quell insubordination.

The Emotion of Power.

The state named the feeling or emotion of Power expresses a first-class motive of the human mind. It is, however, shown, with great probability, not to be an independent source of emotion. It very often consists of a direct reference to possessions or worldly abundance. In other cases, I cannot doubt that the pleasure of malevolent infliction is an element. The love of domineering, or subjecting other people's wills, would be much less attractive than it is, if malevolent possibilities were wholly left out.

Power in the actual is given by bodily and mental superiority, by wealth, and by offices of command

Hence it can be enjoyed in any high degree only by a few. It is, however, capable of great ideal expansion; we can derive gratification from the contemplation of superior power, and the outlets for this are numerous, including not merely the operations of living beings, but the forces of inanimate nature. For example, the Sublime is an ideal of great might or power.

We have now almost, but not quite, led up to the much-urged educational motive, the gratification of the sense of self-activity in the pupils. This must afterwards undergo a very searching examination. Let us, how ever, first briefly review another leading class of well-marked feelings, those designated by the familiar terms —Self-complacency, Pride, Vanity, Love of Applause. Whether these be simple or compound in their nature, they represent feelings of great intensity, and they are specially invoked in the sphere of education.

The Emotions of Self.

'Self' is a very wide word. 'Selfish,' 'Self-seeking,' 'Self-love,' might be employed without bringing any new emotions to the front. All the sources of pleasure, and all the exemptions from pain, that have been or might be enumerated, under the Senses and the Emotions, being totalized, could be designated as 'Self' or 'Self-interest.' But connected with the terms Self-esteem, Self-complacency, Pride, Vanity, Love of Praise, there are new varieties of feeling, albeit they are but offshoots from some of those already given. It is not our business to trace the precise derivation of these complex modes, except to aid in estimating their value as a distinct class of motives.

There is an undoubted pleasure in finding in ourselves some of those qualities that, seen in other men, call forth our love, admiration, reverence, or esteem. The names self-complacency, self-gratulation, self-esteem, indicate emotions of no little force. They have a good influence in promoting the attainment of excellence; their defect is ascribable to our enormous self-partiality: for which cause they are usually concealed from the jealous gaze of our fellows. It is only on very special occasions that persuasion is made to operate through these powerful feelings; they are too ready to turn round and make demands that cannot be complied with.

A still higher form of self-reflected sentiment is that designated by the Love of Praise and Admiration. We necessarily feel an enhanced delight when our own good opinion of self is echoed and sustained by the expressions of others. This is one of the most stirring influences that man can exert over man. It exists in many gradations, according to our love, regard, or admiration for the persons bestowing it, as well as our dependence upon them, and according to the number joining in the tribute.

The bestowal of praise is an act of justice to real merit, and should take place apart from ulterior considerations. But in rewarding, as in punishing, we cannot help looking beyond the present; we have in our eye merits that are yet to be achieved. The fame that attends intellectual eminence is an incentive to study, and the educator has this great instrument at his command.

Praise, to be effectual and safe, has to be carefully apportioned, so as to approve itself to all concerned. As the act of praising does not terminate with the moment, but establishes claims for the future, thoughtless profusion of compliment defeats itself. Praise may operate in the form of warm kindly expression, and no more ; in which sense it is an offering of affection, and has a value in that character alone. A pleased smile is a moral influence.

Discipline, properly so called, works in the direction of pain ; pleasures are therein viewed in their painful obverse. The positive value of delights is of consequence only as the starting-point wherefrom to count the efficacy of deprivations. The pains opposed to the pleasures of Self-esteem and Praise are among the most powerful weapons in the armoury of the disciplinarian. They are the chief reliance of such as deprecate corporal inflictions. Bentham's elaborate scheme of discipline in the ' Chrestomathia ' is a manipulation of the motives of Praise and Dispraise, which he would fain make us believe to be all-sufficient.

Of the two divisions of the present class of emotions, namely, Self-esteem on the one hand, and Desire of Praise on the other, the opposite of the first—Self-reproach, Self-humbling—is very little under foreign influence. To induce people to think meanly of themselves is no easy task ; with the mass of human beings it is well-nigh hopeless. Any success that attends the endeavour is to be traced to the second member of the class under discussion, namely, Dispraise, Depreciation. There is no mistaking our aim here ; we can make our power felt in this form, whether it has the other effect or not. People live so much on one another's good opinion that the remission tells in an instant ; from the simple abatement or loss of estimation there is a descent into the depths of disesteem with a result of unspeakable suffering. The efforts that the victim makes to right himself under censure only show how keenly it is felt. There can be little doubt that on the delicate handling of this instrument must depend the highest refinements of moral control.

The Emotions of Intellect.

The pleasurable emotions incident to the exercise of the Intellectual Powers have not the formidable magnitude that we have assigned to the foregoing groups. Indeed, even on the occasions when they seem to burst forth with an intense glow, we can discern the presence of emanations from these other great fountains of feeling.

It is an effort of prime importance to trace exhaustively the inducements and allurements to intellectual exertion. What are the intrinsic charms of knowledge, whether in pursuit or in possession? The difficulty of the answer is increased, rather than diminished, by the flow of fifty years' rhetoric.

Knowledge has such a wide compass, embraces such various ingredients, that, until we discriminate the kinds of it, we cannot speak precisely either of the charms it possesses or of the absence of charm. Some sorts of knowledge are interesting to everybody ; some interest only a few. The serious part of the case is, that the most valuable kinds of knowledge are often the least interesting.

The important distinction to be drawn here is between Individual or Concrete Knowledge, and General or Abstract Knowledge. As a rule, particulars are interesting as well as easy; generals uninteresting and hard. When particulars are not interesting, it is often from their being overshadowed by generals. When generals are made interesting, it is by a happy reflected influence upon the particulars. It would serve nearly· all the purposes of the teacher to know the best means of overcoming the repugnance and the abstruseness of general knowledge.

Waiving for a time the niceties of the abstract idea, and the obstacles in the way of its being readily comprehended, we may here adduce certain motives that cooperate with the teacher's endeavours to impress it. A little attention, however, must first be given to the various kinds of interest that pertain to individual or particular facts.

Any kind of knowledge, whether particular or more or less general, that is obviously involved in any of the strong feelings or emotions that we have passed in review, is by that very fact interesting. Now a great many kinds of knowledge are implicated with those various feelings. To avoid pains, and obtain pleasures, it is often necessary to know certain things, and we willingly apply our minds to learn those things; and the more so, the more evident their bearing upon the gratification of our desires. A vast quantity of information respecting the world, and respecting human beings, is gained in this way; and it constitutes an important basis of even the highest acquisitions.

The readiness to imbibe this immediately fructifying knowledge is qualified by its being difficult or abstruse; we often prefer ignorance, even in matters of consequence, to intellectual labour.

All the natural objects that bear upon our subsistence, our wants, our pleasures, our exemptions from pain, are individually interesting to us, and become known in respect of their special efficacy. Our food, and all the means of procuring it, our clothing and shelter, our means of protection, our sense-stimulants, are studied with avidity, and remembered with ease. This department of knowledge, notwithstanding its vital concern, is apt to be considered as grovelling; it has, however, the recommendation of truth. We do not encourage ourselves in any deceptions in such matters; and, if we make mistakes, it is owing to the obscurity of the case, rather than to our indifference, or to any motive for perverting the facts. Indeed, this is the department that first supplied to mankind the best criterion of certainty.

There is a different class of objects that appeal, not to the more pressing utilities of subsistence, safety, and comfort, but to the gratifications of the higher senses and the emotions: the pleasures of touch, sight and hearing; the social and anti-social emotions. These comprise all the more striking objects of the world :—the sun and the celestial sphere, the earth's gay colouring, sublime vastness, the innumerable objects—inanimate and animate—that tickle some sense or emotion. In proportion as human beings are set free from the struggle for subsistence, do they lay themselves open to these influences, and so enlarge the sphere of natural knowledge. Individual things become interesting and known, from inspiring these feelings. The culminating interest, however, is in living beings, and especially persons of our own species. The intellectual impressions thus left upon us are lively, but not necessarily correct to the facts.

However all this may be, it is to individual things that we must refer the first beginnings of knowledge, the interest and the facility of acquisition. There are great inequalities in this interest and consequent facility. Many individual objects inspire no interest at all in the first instance ; while some of them become interesting afterwards, in consequence of our discovering in them relationships to things of interest.

One notable distinction among the objects of know-ledge is the distinction between movement or change, and stillness or inaction. It is movement that excites us most ; still life is rendered interesting by reference to movement. We are aroused and engrossed by all moving things ; our attention is turned away from ob-jects at rest to contemplate movements ; and we imbibe with great rapidity the impressions of moving objects.

This brief survey of the sphere of Individuality and of the various attractions presented by individuals is preparatory to the consideration of the most arduous part of knowledge—the knowledge of generals or Gene-rality. All the difficulties of the higher knowledge have reference to the generalizing process—the seeing of one in many. The arts of the teacher and the expositor are supremely requisite in sweetening the toil of this opera-tion. At the present stage, however, the question is to assign the motives connected with general knowledge as distinct from individual knowledge.

General knowledge, represented by Science, consists in holding together, by a single grasp, whole classes of objects, of facts, of operations. This must, by the very nature of the case, be more severe than holding an in-dividual. To form an idea of one tree that we have repeatedly surveyed at leisure round and round, is about the easiest exertion whether of attention or of memory. To form an idea of ten trees partly agreeing and partly differing among themselves, is manifestly an entirely altered task; it is to exchange comparative simplicity for arduous complexity: yet this is what is needed everywhere in the higher knowledge.

The first emotional effect attendant on the process of generalizing facts, and serving to lighten the intellectual burden, is the flash of identity in diversity; an exhilarating charm that has been felt in every age by the searchers after truth. Many of the grandest discoveries in science have consisted, not in bringing to light any new individual fact, but in seeing a likeness between things formerly regarded as wholly unlike. Such was the great discovery of gravitation. The first flash of the recognition of a common power in the motions of the planets and the flight of a projectile on the earth was unutterably splendid; and after a hundred repetitions, the emotional charm is unexhausted.

With the emotion of exhilarating surprise at the discovery of likeness among things seemingly unlike, there is another grateful feeling—the relief from an intellectual burden. This appears at first sight a contradiction to what has been already said respecting the greater laboriousness of general knowledge; but the contrariety is only apparent. To contract an impression of one single individual, after plenty of time given to attend to it, is the easiest supposable mental effort. But such is the multiplicity of things, that we must learn to know, and remember, vast numbers of individuals; and we soon feel ourselves overpowered by the never-ending demands upon us. We must know many persons, many places, many houses, many natural objects; and our capability of memory is in danger of exhaustion before we have done. Now, however, comes in the discovery of identities, by which the work is shortened. If a new individual is exactly the same as the old, we are saved the labour of a new impression ; if there is a slight difference, we have to learn that difference and no more. In actual experience, the case is, that there are numerous agreements in the world, but accompanied with differences ; and while we have the benefit of the agreements, we must take notice of the differences. What makes a general notion difficult is that it represents a large number of objects that, while agreeing in some respects, differ in others. This difficulty is the price that we pay for an enormous saving in intellectual labour.

The overcoming of isolation in the multitude of particulars, by flashes of identity, is the progress of our knowledge in one direction ; it is the satisfaction that we express when we say we understand or can account . for a thing. Lightning was accounted for when it was identified with the electric spark. Besides the exhilarating surprise at the sameness of two facts in their nature so different and remote, men had the further satisfaction of saying that they learned what lightning is. Thus by discoveries of identity we are enabled to explain the world, to assign the causes of things, to dissipate in part the mysteriousness that everywhere surrounds us.

When a discovery of identification is made among particulars hitherto looked upon as diverse, the interest created is all-sufficient to secure our appreciation. This

is the alluring side of generalities. The repugnant aspect
of them is seen in the technical language invented to
hold and express them—general or abstract designations,
diagrams, formulas. When it is proposed to indoc-
trinate the mind in these things, by themselves, and
at a stage when the condensing and explaining power
of the identities is as yet unawakened, the whole machinery seems an uncouth jargon. Hence the attempt to
afford relief to the faculties by teaching the dry symbols
of Arithmetic and Geometry through the aid of examples
in the concrete, and, in all the abstract sciences, to afford
plenty of particulars to illustrate the generalities. This
is good so far ; but the real interest that overcomes the
dryness arises only when we can apply the generalities
in tracing identities, in solving difficulties, and in short-
ening labour ; an effect that comes soonest to those
that have already some familiarity with the field where
the formulas are applicable. The liking for Algebra and
for Geometry proceeds apace when one sees the marvels
of curious problems solved, unlikely properties discovered,
among numbers and geometrical figures. A certain ease
in holding in the memory the abstract symbols, after a
moderate application, is enough to prepare us for a positive relish in the pursuit. Such is the case with generalities in all departments. If we can hold on till they
bear their fruits in the explanation of things that we
have already begun to take notice of, the pursuit is
sustained by a genuine and proper scientific interest,
whose real groundwork, however deeply hidden, is the
stimulus of agreement among differing particulars, and
the lightening of the intellectual labour in comprehending the world. These are the feelings that have to be

8

awakened in the minds of pupils when groaning under the burden of abstractions.

The opposition of the Concrete and the Abstract, while but another way of expressing the opposition of the Particular and the General, brings into greater prominence the highly *composite* or combined character of Individuality. The individual thing is usually a compound of many qualities, each of which has to be abstracted in turn, in rising to general notions: any individual ball has, in addition to its round form, the properties called weight, hardness, colour, and so on. Now this composite nature, by charming several senses at once, gives a greater interest to individuals, and urges us to resist that process of decomposition, and separate attention, known by the designations 'abstraction' and 'analysis.' It is for individuals in all their multiplicity of influence that we contract likings or affections; and in proportion as the charm of sense, and especially the colour sense, is strong in us, we are averse to the classing or generalizing operation. A fire is an object of strong individual interest: to rise from this to the general notion of the oxidation of carbon under all varieties of mode, including cases with no intrinsic charm, is to quit with reluctance an agreeable contemplation. The emotions now described—the pleasure of identity, and the lightening of labour—are of avail to counter-work this reluctance.

The second of the two motives that we have coupled together—the easing of intellectual labour—may be viewed in another light. When objects are regarded as operating agents in the economy of the world, as causes or instruments of change, they work by their qualities

or powers in separation, and not by their entire individuality or concreteness. An iron bar, or a poker, is an
individual concrete thing ; but when we come to use it,
we put in action its various qualities separately. We
may employ it as a weight ; in which case its other properties are of no account. We may use it as a lever, and
then we bring into play simply its length and its tenacity.
We can put it in motion as a moving power, when its
inertia alone is taken into account, with perhaps its form.
In all these instances, the magnetical and the chemical
and the medicinal properties of iron are unthought of.
Now this consideration discloses an important aid to the
abstracting process—the analytic separation of properties,
as opposed to the mind's fondness for clinging to concrete individuality. When we are working out practical
ends, we must follow nature's method of working ; and,
as that is by isolating the separate qualities, we must
perform the act of mental isolation, which is to abstract,
or consider one power to the neglect of the rest. When
we want to put forth heavy pressure, we think of various
bodies solely as they can exert weight, however many
other ways they may invite or charm our sense. This
is to generalize or form a general notion of weight ;
and the motive to conceive it, is practical need or necessity.

This motive of practical need at once brings us to
the very core of Causation, viewed as a merely speculative notion. The cause of anything is the agent that
would bring that thing into being, suppose we were in
want of it. The cause of warmth in a room is combustion properly arranged. We use this fact for practical
purposes, and we may also use it for satisfying mere

curiosity. We enter a warm room ; we may desire to know how it has been made warm, and we are satisfied by being told that there has been, or is now somewhere, a fire in communication with it.

Thus it is that in proportion as we come to operate upon the world practically ourselves, and from that pro-ceed to contemplate causation at large, we are driven upon the abstracting and analyzing process, so repugnant to one large portion of our feelings. Science finds an opening in our minds at this point, when otherwise we might need the proverbial surgical operation.

These observations will serve to illustrate the working of the emotion named Curiosity, which is justly held to be a great power in teaching. Curiosity expresses the emotions of knowledge viewed as desire ; and, more especially, the desire to surmount an intellectual difficulty once felt. Genuine curiosity belongs to the stage of advanced and correct views of the world.

Much of the curiosity of children, and of others besides children, is a spurious article. Frequently it is a mere display of egotism, the delight in giving trouble, in being pandered to and served. Questions are put, not from the desire of rational information, but from the love of excitement. Occasionally, the inquisitiveness of a child provides an opportunity for imparting a piece of real information ; but far oftener not. By ingeniously circumventing a scientific fact, one not too high for a child's comprehension, we may awaken curiosity and succeed in impressing the fact. Try a child to lift a heavy weight first by the direct pull, and then by a lever or a set of pulleys, and probably you will excite

some surprise and wonder, with a desire to know something further about the instrumentality. But one fatal
defect of the childish mind is the ascendancy of the
personal or anthropomorphic conception of cause. This
no doubt is favourable to the theological explanation of
the world, but wholly unsuited to physical science. A
child, if it had any curiosity at all, would like to know
what makes the grass grow, the rain fall, the wind howl,
and generally all things that are occasional and exceptional ; an indifference being contracted towards what
is familiar, constant, and regular. When anything goes
wrong, the child has the wish to set it right, and is
anxious to know what will answer the purpose : this is
the inlet of practice, and, by this, correct knowledge
may find its way to the mind, provided the power of
comprehension is sufficiently matured. Still, the radical obstacle remains—the impossibility of approaching
science at random, or taking it in any order ; we must
begin at the proper beginning, and we may not always
contrive to tickle the curiosity at the exact stage of the
pupil's understanding. Every teacher knows, or should
know, the little arts of giving a touch of wonder and
mystery to a fact before giving the explanation ; all
which is found to tell in the regular march of exposition, but would be lost labour in any other course.

The very young, those that we are working upon by
gentle allurement, are not fully competent to learn the
'how' or the 'wherefore' of any important natural fact;
they cannot even be made to desire the thing in the
proper way. They are open chiefly to the charm of
sense, novelty, and variety, which, together with accidental charm or liking, impresses the pictorial or concrete

aspects of the world, whether quiescent or changing, the last being the most powerful.   They farther are capable of understanding the more palpable conditions of many changes, without penetrating to ultimate causes.   They learn that to light a fire there must be fuel and a light applied ; that the growth of vegetables needs planting or sowing, together with rain and sunshine through a summer season.   The empirical knowledge of the world that preceded science is still the knowledge that the child passes through in the way to science ; and all this may be guided so as to prepare for the future scientific revelations.   In other respects, the so-called curiosity of children is chiefly valuable as yielding ludicrous situations for our comic literature.

The Emotions of Activity.

Nothing is more frequently prescribed in education than to foster the pupils' own activity, to put them in the way of discovering facts and principles for themselves.   This position needs to be carefully surveyed.

There is, in the human system, a certain spontaneity of action, the result of central energy, independent of any feelings that may accompany the exercise.   It is great in children ; and it marks special individuals, who are said to possess the active temperament.   It distinguishes races and nationalities of human beings, and is illustrated in the differences among the animal tribes ; it also varies with general bodily vigour.   This activity would burst out and discharge itself in some form of exertion, whether useful or useless, even if the result were perfectly indifferent as regards pleasure or pain

We usually endeavour to turn it to account by giving it a profitable direction, instead of letting it run to waste or something worse. It expends itself in a longer or a shorter time, but, while any portion of it remains, exertion is not burdensome.

Although the spontaneous flow of activity is best displayed and is most intelligible in the department of muscular exercise, it applies also to the senses and the nerves, and comprises mental action as well as bodily. The intellectual strain of attention, of volition, of memory, and of thought, proceeds to a certain length by mere fulness of power, after rest and renovation; and may be counted on to this extent as involving nothing essentially toilsome. Here, too, a good direction is all that is wanted to make a profitable result.

The activity thus assumed as independent of feeling is nevertheless accompanied with feeling, and that feeling is essentially pleasurable : the pleasure being greatest at first. The presence of pleasure is the standing motive to action ; and all the natural activity of the system — whether muscular or nervous — brings an effluence of pleasure, until a certain point of depletion is arrived at.

If, further, our activity is employed productively, or in yielding any gratification beyond the mere exercise, this is so much added to the pleasures of action. When, besides the delight of intellectual exercise, we obtain for ourselves the gratification of fresh knowledge, we seem to attain the full pleasure due to the employment of the intellect.

Much more, however, is meant by the gratification of the self-activity of the learner. That expression

points to the acquiring of knowledge, as little as pos·
sible by direct communication, and as much as possible
by the mind's own exertion in working it out from the
raw materials. We are to place the pupil as nearly as
may be in the track of the first discoverer, and thus
impart the stimulus of invention, with the accompanying
outburst of self-gratulation and triumph. This bold
fiction is sometimes put forward as one of the regular
arts of the teacher; but I should prefer to consider it
as an extraordinary device, admissible only on special
.occasions.

It is an obvious defect in teaching to keep continually lecturing pupils, without asking them in turn to
reproduce and apply what is said. This is no doubt
a sin against the pupil's self-activity, but rather in the
manner than in the fact. Listening and imbibing constitute a mode of activity ; only it may be overdone in
being out of proportion to the other exercises requisite
for fixing our knowledge. When these other activities
are fairly plied, the pupil may have a certain complacent
satisfaction in his or her own efficiency as a learner,
and this is a fair and legitimate reward to an apt pupil.
It does not assume any independent self-sufficiency ; it
merely supposes an adequate comprehension and a
faithful reproduction of the knowledge communicated.
The praise or approbation of the master, and of others
interested, is a superadded reward.

Notwithstanding, there still remains, if we could
command it, a tenfold power in the feeling of origina·
tion, invention, or creation; but as this can hardly ever
be actual, the suggestion is to give it in fiction or imagi·
·nation. Now, it is one of the delicate arts of an accom·

plished instructor, to lay before the pupils a set of facts
pointing to a conclusion, and to leave them to draw the
conclusion for themselves. Exactly to hit the mean
between a leap too small to have any merit, and one too
wide for the ordinary pupil, is a fine adjustment and a
great success. All this, however, belongs to the occasional luxuries, the bon-bons of teaching, and cannot be
included under the daily routine.

It is to be borne in mind that although the pride of
origination is a motive of extraordinary power, and in
some minds surpasses every other motive, and has a
great charm even in a fictitious example, yet it is not in
all minds the only extraneous motive that may aid the
teacher. There is a counter motive of sympathy, affection, and admiration for superior wisdom, which operates
in the other direction ; giving a zest in receiving and
imbibing to the letter what is imparted, and jealously
restraining any independent exercise of judgment such
as would share the credit with the instructor. This
tendency is no doubt liable to run into slavishness and
to favour the perpetuation of error and the stagnation
of the human mind ; but a certain measure of it is only
becoming the attitude of a learner. It accompanies a
proper sense of what is the fact, namely, that the
learner *is* a learner and not a teacher or a discoverer,
and has to receive a great deal with mere passive
acquiescence, before venturing to suggest any improvements. Unreasoning blind faith is indispensable in
beginning any art or science ; the pupil has to lay up a
stock of notions before having any materials for discovery or origination. There is a right moment for
relaxing this attitude, and for assuming the exercise of

independence; but it has scarcely arrived while the schoolmaster is still at work. Even in the higher walks of university teaching, independence is premature, unless in some exceptional minds, and the attempt of masters to proceed upon it, and to invite the free criticism of pupils, does not appear ever to have been very fruitful.[1]

The Emotions of Fine Art.

This is necessarily a wide subject, but for our purpose a few select points will be enough. The proper and principal end of Art is enjoyment. Now, whatever is able to contribute on the great scale to our enjoyment, is a power over all that we do. The bearings of this on education are to be seen.

The Art Emotions are seldom looked upon as a mere source of enjoyment. They are apt to be regarded in preference as a moral power, and an aid to education at every point. Nevertheless, we should commence with

---

[1] It would lead us too far, although it might not be uninstructive, to reflect upon the evil side of this fondness for giving a new and self-suggested cast to all received knowledge. It introduces change for the mere sake of change and never lets well alone. It multiplies variations of form and phraseology for expressing the same facts, and so renders all subjects more perplexed than they need be; not to speak of controverting what is established, because it is established, and allowing nothing ever to settle. Owing to a dread of the feverish love of change, certain works that have accidentally received an ascendancy, such as the Elements of Euclid, are retained notwithstanding their imperfections. The acquiescent multitude of minds regard this as a less evil than letting loose the men of action and revolution to vie with each other in distracting alterations, while there is no judicial power to hold the balance. It is a received maxim in the tactics of legislation that no scheme, however well matured, can pass a popular body without amendment; it is not in collective human nature to accept anything *simpliciter*, without having a finger in the pie.

recognizing in them a means of pleasure as such, a pure hedonic factor; in which capacity they are a final end. Their function in intellectual education is the function of all pleasure when not too great, namely, to cheer, refresh, and encourage us in our work.

There are certain general effects of Art that come in well at the very beginning. Such are symmetry, order, rhythm, and simple design and proportion ; which are the adjuncts of the school, just as they should be the adjuncts of home life. Proportion, simple design, a certain amount of colour, are the suitable elements of the school interior ; to which are added tidiness, neat-ness, and arrangement, among the pupils themselves; only this must not be worrying and oppressive.

In the exercises suited to infants, Time and Rhythm are largely employed.

Of all the fine arts, the most available, universal and influential is Music. This is perhaps the most unexcep-tionable, as well as the cheapest, of human pleasures. It has been seized upon with avidity by the human race in all times ; so much so, that we wonder how life could ever have been passed without it. In the earlier stages, it was united with Poetry, and the poetical element was of equal, if not of greater, power than the musical accompaniment. As the ethical instructors of mankind have always disavowed the pursuit of pleasure as such, and allowed it only as subsidiary to morality and social duty, the question with legislators has been what form of music is best calculated to educe the moral virtues and the nobler characteristics of the mind. It was this view that entered into the speculative social constructions of Plato and of Aristotle. Now, undoubtedly the various modes of music operate very differently on the mind ; everyone knows the extremes of martial and ecclesiastical music ; and fancy can insert many intermediate grades.[1]

For the moment a musical strain exerts immense power over the mind, to animate, to encourage, to soothe, and to console. But the facts do not bear us out in attributing to it any permanent moral influence ; nothing is more fugitive than the excitement of a musical performance. Excepting its value as a substantive contribution to the enjoyment of life, I am not able to affirm that it has any influence on education, whether moral or intellectual. Certainly, if it has any effect in the moral sphere, it has none that I can trace in the sphere of intellect. As a recreative variety in the midst of toil, it deserves every encomium. In those exercises that are half recreative, half educational, as drill and gymnastics, the accompaniment of a band is most stimulating. In the Kindergarten it is well brought in, as the wind-up to the morning's work. But music during ordinary lessons, or during any sort of intellectual work, is mere distraction, as everyone knows from the experience of street bands and organs.

Excess in the pleasures of music, like every other excess, is unfavourable to mental culture. But some of the most intellectual men that ever lived have been de-

[1] Plato, in the *Republic*, wishing to train a vigorous and hardy race, interdicted not simply the unfavourable musical strains, but the instruments most adapted to these. He permits only the lyre and the harp, with the panspipe for shepherds attending their flocks ; forbidding both the flute and all complicated stringed instruments. Disallowing the lugubrious, passionate, soft, and convivial modes of music, he tolerates none but the Dorian and the Phrygian, suitable to a sober, resolute, courageous frame of mind ; to which also the rhythm and movement of the body is to be adapted. (Grote's *Plato*, III. 196.)

votees of music. In the case of Luther, it seems to have been incorporated with his whole being ; Milton invoked it as an aid in poetic inspiration. These were men whose genius largely involved their emotions. But the musical enthusiasm of Jeremy Bentham could have no bearing on his work, further than as so much enjoyment.

Poetry is music and a great deal more. Its bearings are more numerous and complicated. In the ruder stages of music, when it accompanied poetry, the main effects lay in the poetry. The poetic form—the rhythm and the metre—impresses the ear, and is an aid to memory ; whence it has been transferred from the proper themes of poetry to very prosaic subjects by way of a mnemonic device. The subject-matter of poetry comprises the stirring narrative, which is an enormous power in human life, and the earliest intellectual stimulus in education.

The Ethical Emotions.

The feelings called Ethical, or Moral, from their very meaning, are the support of all good and right conduct. The other emotions may be made to point to this end, but they may also work in the opposite direction.

When the educator describes these in more precise and equivalent phraseology, he generally singles out regard to the pleasure and displeasure of parents and superiors, together with habits or dispositions towards obedience ; all which is the result of culture and growth.

Any primitive feelings conspiring towards good conduct must be of the nature of the sympathies or social yearnings ; which are called into exercise in definite ways, well known to all students of human nature. By far the most powerful stimulus to acts of goodness towards others, is good conduct on the part of others : whoever can resist this is a fit subject for the government of fear, and nothing else. The law says, ' Do unto others as ye *would* that they should do unto you.' The lower ground of practice is, ' Do unto others as they do unto you.' This is as far as the very young can reach in moral virtue.

It is too much to expect in early years generous and disinterested impulses, unreciprocated. The young have little to call their own ; they have no means. Their fortune is their free, unrestrained vivacity, their elation, and their hopes. If they freely give up any part of this, it is in consideration of equivalent benefits. They are susceptible of being worked up to moments of self-renunciation ; in which they may commit their future irrevocably, without knowing what they are about. But they cannot be counted on for daily, persistent self-restraint, willingly encountered, unless there be some seen reward, present or in the distance. It takes a good deal to bring anyone even up to the point of rendering *quid pro quo* in all things.

The Feelings as Appealed to in Discipline.

The survey that has now been made of the sensibilities of the human mind, available as motives, prepares for the consideration of Discipline in teaching. The instructor finds that, in school moments, and for school purposes, he has to restrain all the unruly impulses, and to overbear the sluggishness of the youthful nature. To succeed in this requirement many arts are employed, corresponding to the wide compass of sensations and emotions that agitate the human breast.

The question how to maintain discipline among masses of human beings is of very wide application, and is therefore the subject of a great variety of experiments In the wide field of moral control, it includes a principal function of government, namely, the repression of crime : a department that has lately received much attention. To collect all the lights furnished in each of the spheres where moral control has to be exercised, is to contribute to the illumination of each.   There has, undoubtedly, in former times, been very great mismanagement in almost every one of the regions of repressive authority—in the state, in the family, and in the school ; in all which an excess of human misery has been habitually engendered by badness in the manner of exercising control.   It is, perhaps, in the family that the mischief is most widely spread and most baneful.

By degrees we have become aware of various errors that ran through the former methods of discipline, in the several institutions of the state, as well as in the family. We have discovered the evil of working by fear alone, and still more by fear of coarse, painful, and degrading inflictions.   We have discovered that occasions of offence can be avoided by a variety of salutary arrangements, such as to check the very disposition to unruly conduct. We consider that a great discovery has been made in regard to punishments, by the enunciation of the maxim that certainty is more important than severity ; to which should be added, proportion to the offence.   We also consider that by a suitable training, or education, the dispositions that lead to disorder and crime can be

checked in the bud ; and that, until there has been room
for such training to operate, the mind should not be
exposed to temptation.   We have become accustomed
to lay more stress than was formerly done upon cultivating the amicable relations of human beings ; the
tendency of which is to abridge the sphere of injurious
conduct on the part of individuals.

The consideration of discipline in Education supposes
the relation of a teacher to a class ; one man or woman
exercising over a body of pupils the authority requisite
for the work in hand.   Nevertheless, it is not lost time
to advert, in the first instance, to the maxims pertaining
to authority in general.

Authority, government, power over others is not an
end in itself ; it is only a means.   Further, its operation
is an evil ; it seriously abates human happiness.   The
restraint upon free agency, the infliction of pain on individuals, the setting up a reign of terror—all this is
justified solely by the prevention of evils out of all proportion to the misery that it inflicts.   This might seem
self-evident ; but is not so.   The deep-seated malevolence and lust of domination in the human mind
make the necessity of government a pretext for excesses
in severity and repression ; to which must be added
the opportunity of preying upon the substance of the
governed.

The philosophy of society now endeavours to formu-
late the limits to authority, and to the employment of
repressive severities.   Not only is authority restricted
to the mildest penalties that will answer its purpose ;
but its very existence has to be justified in each case
that arises.

Authority is not necessary to every teaching relation. A willing pupil coming up to a master to be taught, is not entering into a relationship of authority ; it is a mere voluntary compact, terminable at the pleasure of each. There is no more authority over the assemblies of grown men to hear lectures, than over the worshippers at church, or the frequenters of the play.   There is nothing but the observance of mutual toleration and forbearance so far as is requisite to the common good ; if this were grossly violated, there would be an exercise of power either by the collective mass themselves, or by summoning the constable to their aid.   No authority is lodged in the lecturer, the preacher, or the performer, to repress disturbances.

Authority first appears in the family, and is thence transferred with modifications to the school.   It is between these two institutions that the comparison is most suggestive.   The parent's authority is associated with sustenance, and has an almost unlimited range.   It is tempered by affection, but this depends upon mutuality of pleasure-giving, and supposes a limited number. The teacher's authority has nothing to do with sustenance, it is a duty undertaken for payment ; it is subsidiary to the single object of teaching a definite amount of knowledge.   It wants the requisites of affection ; the numbers are too great, and the mutual concern too restricted : yet affection is not wholly excluded ; in certain well-marked cases, it may play a part.

On the other hand, the family and the school have some important agreements.   They both deal with immature minds, for whom certain kinds of motives are

9

unsuitable. Neither can employ motives that are applicable only to grown men and women; they cannot appeal to consequences in the distant and unknown future. Children do not realize a remote effect, and they fail even to conceive many things that will one day have great power over their conduct. To talk to them about riches, honours, and a good conscience is in vain. A half-holiday is more to them than the prospect of becoming head of a business.

The position of immaturity is attended with another peculiarity, namely, that the reasons of a rule cannot always be made apparent. Sometimes they can, if not to the younger, at least to the older children. This is a highly prized aid to obedience in every department of government.

There are many important points of agreement in the exercise of authority in every sphere—the family, the school, the relation of master and servant, of ruler and subject, whether in the state at large or in subordinate societies. For example :—

(1) Restraints should be as few as the situation admits of.

(2) Duties and Offences should be definitely expressed, so as to be clearly understood. This may not always be possible to the full extent; but should be always aimed at.

(3) Offences should be graduated according to their degree of heinousness. This too needs clearness of discrimination and definite language.

(4) The application of Punishment is regulated according to certain principles, first clearly pointed out by Bentham.

(5) Voluntary dispositions are to be trusted as far as they can go.

(6) By organization and arrangement, the occasions of disorder are avoided. Quarrels are obviated by not permitting crowds, jostling, and collisions. Dishonesty is checked by want of opportunity; remissness, by the watchful eye and by definite tests of performance.

(7) The awe and influence of authority is maintained by a certain formality and state. Forms and ritual are adapted to all the operations of law : persons in authority are clothed with dignity and inviolability. The greater the necessity of enforcing obedience, the more stern and imposing is the ritual of authority. The Romans, the greatest law-giving people, were the most stately in their official rites. A slight tinge of formality should accompany even the lowest forms of authority.

(8) It is understood that authority, with all its appurtenances, exists for the benefit of the governed, and not as a perquisite of the governor.

(9) The operation of mere vindictiveness should be curtailed to the uttermost.

(10) So far as circumstances allow, everyone in authority should assume a benign character, seeking the benefit of those under him, using instruction and moral suasion so as to stave off the necessity of force. The effect of this attitude is at its utmost when its limits are clearly discerned and never passed.

(11) The reasons for repression and discipline should, as far as possible, be made intelligible to those concerned ; and should be referable solely to the general good. This involves, as a part of national education, a

knowledge of the structure of society, as being a regulated reciprocity among all its members, for the good of
each and of all.[1]

[1] Whoever occupies a position of authority ought to be familiar with the
general principles and conditions of Punishment, as they may be found set
forth in the Penal Code of Bentham.   The broad, exhaustive view there
given will co-operate beneficially with each one's actual experience.   I
make no apology for presenting a short summary of his principles.

After precisely defining the proper ends of Punishment, Bentham marks
the cases unmeet for Punishment.   First, where it is *groundless* : that is,
where there never has been any real mischief (the other party consenting to
what has been done), or where the mischief is overweighed by a benefit of
greater value.   Second, where it is *inefficacious* : including cases where
the penal provision has not come before the offender's notice, where he is
unaware of the consequences of his act, or where he is not a free agent.
Third, cases where it is *unprofitable* : that is, when the evil of the punishment exceeds the evil of the offence.   (The evils of Punishment, which
have to be summed up and set against the good, are (1) coercion or restraint, (2) the uneasiness of apprehension, (3) the actual suffering, (4) the
suffering caused to all those that are in sympathy with the person punished.)
Fourth, cases where Punishment is *needless* : as when the end can be
attained in some cheaper way, as by instruction and persuasion.   In this
class Bentham specially includes the offences that consist in disseminat-
ing pernicious principles in politics, morality, or religion.   These should
be met by instruction and argument, and not by the penalties of the law.

Under what he calls the expense or frugality of Punishment, Bentham
urges the necessity of presenting to the mind an adequate notion of what a
punishment really is.   Hence the advantage of punishments that are easily
learnt, and remembered, and that appear greater, and not less, than they
really are. .

Next, as to the main point, the *measure* of Punishment.   First, it should
be such as clearly to outweigh the profit of the offence : including not
simply the immediate profit, but every advantage, real or apparent, that
has weighed as an inducement to commit it.   Second, the greater the
mischief of the offence, the greater is the expense that it is worth while to
be at, in the way of punishment.   Third, when two offences come into
competition, the punishment for the greater should be such as to make the
less preferred ; thus robbery with violence to the person, is always punished
more severely than simple robbery.   Fourth, the punishment to be so
adjusted that for every part of the resulting mischief a motive may be
provided to restrain from causing it.   Fifth, the punishment should not be

The points of comparison and of contrast between
the school and the family have been noted. The more

greater than is needed for these ends. Sixth, there should be taken into
account the circumstances affecting the sensibility of the offenders, as age,
sex, wealth, position, so that the same punishment may not operate unequally. Seventh, the punishment needs to be increased in magnitude as
i falls short of certainty. Eighth, it must be further increased in magnitude as it falls short in point of proximity. Penalties that are uncertain,
and those that are remote, correspondingly fail to influence the mind.
Ninth, when the act indicates a *habit*, the punishment must be increased
so as to outweigh the profit of the other offences that the offender may
commit with impunity : this is severe, but necessary, as in putting down
the coiners of base money. Tenth, when a punishment well fitted in its
quality cannot exist in less than a certain quantity, it may be of use to
employ it, although a little beyond the measure of the offence : such are
the punishments of exile, expulsion from a society, dismissal from office.
Eleventh, this may be the case more particularly, when the punishment is
a moral lesson. Twelfth, in adjusting the quantum, account is to be taken
of the circumstances that render all punishment unprofitable. Thirteenth,
if in carrying out these provisions anything occurs tending to do more
harm than the good arising from the punishment, that thing should be
omitted.

In regard to the *selection* of punishments, Bentham lays down a number
of tests, or conditions whereby they are fitted to comply with the foregoing
requirements. First, is the quality of *variability* : a punishment should
have degrees of intensity and duration : this applies to fines, corporal
punishment, and imprisonment ; also to censure, or ill-name. Second,
equability, or equal application under all circumstances : this is not easy
to secure ; a fixed fine is an unequable punishment. Third, *commen-
surability* : that is, punishments should be so adapted to offences, that the
offender may clearly conceive the inequality of the suffering attached to
crimes of different degrees of heinousness : this property can be grafted on
the variable punishments, as imprisonment. Fourth, *characteristicalness* :
this is where something can be found in the punishment whose idea
exactly fits the crime. Bentham dilates upon this topic, in order to
discriminate it from the old crude method of an eye for an eye : cases in
point occur abundantly both in the family and in the school. Fifth,
exemplarity : this is connected with the impressiveness of a punishment ;
all the solemnities accompanying the execution increase this effect.
Bentham, however, did not sufficiently consider the evils attending too
great publicity, which have led to withdrawing punishments from the gaze

special distinction of the school, as compared with relations of authority in general, is resolvable into its main
object, namely, Instruction, for which the condition
that needs to be imposed is Attention and Application
of mind, with a view to permanent impressions—intellectual and other.   To evoke, charm, cajole, compel this
attitude, is the first aim in all teaching.   The hostile
influences to be overcome are such as physical inability
and exhaustion, irksomeness in the work, diversions and
distractions from other tastes, with the natural rebelliousness of human beings under authority.

The arts of proceeding are not the same for a single
pupil and for a class.   For the single pupil, individuality may be studied and appealed to ; for the class,
individualities are not considered.   Here, the element of
number is an essential feature ; carrying with it both
obstructions and aids, and demanding a very special
manipulation.

It is in dealing with numbers that the teacher stands
distinguished from the parent, and is allied to the wider
authorities of the State ; exercising larger control, en-

of the multitude ; it being now simply intimated that they have been carried
out.   Sixth, *frugality* : or making punishments less costly to the State :
as when prisoners are employed productively.   Seventh, *subserviency to
reformation* : by weakening the seductive, and strengthening the preserving
motives ; as in giving habits of labour to the idle.   Eighth, *efficacy in
disablement* : as in deposition from office.   Ninth, *subserviency to compen-
sation* : as by pecuniary inflictions.   Tenth, *popularity*.   Bentham lays
much stress upon the popularity and unpopularity of punishments, whereby
the public sympathy may work for or against the law : when a punishment
is unpopular, juries are reluctant to convict, and public agitation is stirred
for remission of sentence.   Eleventh, *simplicity of description* : under this
head, Bentham comments upon the obscure and unintelligible descriptions
of the old law, as *capital felony, præmunire*.   Twelfth, *remissibility*, in case
of mistake.

countering greater risks, and requiring a more steady hand. With an individual pupil, we need only such motives as are personal to himself; with numbers, we are under the harsh necessity of punishing for example.

Good physical surroundings are known to be half the battle. A spacious and airy building ; room for the classes to come together and to depart without confusion or collision : these are prime facilities and aids to discipline. Next is organization, or method and orderly arrangement in all the movements ; whereby each pupil is always found in the proper place, and the entire mass is comprehended under the master's glance. To this follows the due alternation and remission of work, avoiding fatigue and maintaining the spirits and the energies while the teaching lasts.

After the externals and arrangements, come the Methods and Arts of Teaching, considered as imparting lucidity to the explanations, and as easing the necessary intellectual labour of comprehension. If to this prime quality can be added extraneous interest or charm, so much the better ; but it must not be at the expense of clearness, which is the first condition of getting through the subject.

The personality of the teacher may be in favour of his influence : a likable exterior, a winning voice and manner, a friendly expression, when relaxing the sternness of authority. This is the side of allurement or attraction. The other side is the stately, imposing, and dignified bearing, by which the master can impersonate authority and be a standing reminder to the evil-disposed of the flock. It is seldom given to one man or woman to display both attitudes in their highest force ; but wherever, and to whatever extent, they can be assumed, they constitute a barrier to disaffection and remissness.

Any prominent displays of swagger and self-conceit operate against the teacher's influence, and incite efforts to take him down. It is possible to temper authority with an unassuming demeanour.

Much of course depends upon *tact* ; meaning by that a lively and wakeful sense of everything that is going on. Disorder is the sure sequel of the teacher's failure in sight or in hearing ; but even with the senses good, there may be absent the watchful employment of them. This is itself a natural incapacity for the work of teaching ; just as an orator is sure to fail if he is slow to discern the signs of the effect that he produces on his audience. A teacher must not merely be sensitive to incipient and marked disorder ; he must read the result of his teaching in the pupils' eyes.

That quietness of manner that comes not of feebleness, but of restraint and collectedness, passing easily into energy when required, is a valuable adjunct to discipline. To be fussy and flurried is to infect the class with the same qualities ; unfavourable alike to repression and to learning.

Any mistake, miscarriage, or false step, on the part of a teacher, is for the moment fatal to his ascendancy. Such things *will* happen, and they render undue assumption all the more perilous.

The stress of the teacher's difficulty lies in the heavings of a mass or multitude. The working of human beings collectively, is wholly distinct from their individual action ; a new set of forces and influences

are generated. One man against a multitude is always
in the post of danger. As units in a mass, every individual displays entirely new characters. The anti-social
or malevolent passion—the delight in gaining a triumph
--which is suppressed in the individual as against a
more powerful individual, is re-ignited and inflamed in
company with others. Whenever a simultaneous charge
is possible the authority of a single person is as nought
in the balance.

It is often said that the teacher should get the collective opinion on his side—should, in short, create a
good class-opinion. It is easier to deserve success in
this than to command it. The fear is that, till the end
of time, the sympathy of numbers will continue to
manifest itself against authority in the school. There
will be occasions when the infection of the mass is a
stronghold of order ; as when the majority are bent on
attending to the work, and are thwarted by a few disturbers of the peace ; or when they have a general
sympathy with their teacher, and merely indulge themselves in rare and exceptional outbursts. While a
teacher's merits may gain for him this position of advantage, more or less, he is never above the risks of an
outbreak, and must be ready for the final resort of
repression by discipline or penalties. He may still
work by soothing applications, gentle and kindly re-
monstrance : he may check the spread of disaffection
by watchful tactics, and by showing that he has the
ringleaders in his eye ; but in the end he must punish.

It is this position of constant preparedness for
disorder, sometimes in isolated individuals, and sometimes in the mass, that demands an air and manner

betokening authority, and carrying with it a certain *hauteur* and distance ; the necessity for which is the stronger as the warring elements are more rife.

The discipline of numbers is impeded by two sorts of pupils ; those that have no natural liking for the subject, and those that are too far behind to understand the teaching. In a perfectly-arranged school, both sorts would be excluded from a class.

The foregoing considerations lead up to the final subject—Punishment ; in administering which, the practice of Education, as well as of other kinds of government, has greatly improved. The general principles of punishment have been already announced. We have to consider their application to the school. But first a few words on the employment of Reward.

Emulation—Prizes—Place-taking.

All these names point to the same fact and the same motive—the desire of surpassing others, of gaining distinction ; a motive that has already been weighed. It is the most powerful known stimulant to intellectual application ; and where it is in full operation, nothing else is needed. Its defects are—(1) it is an anti-social principle ; (2) it is apt to be too energetic ; (3) it is limited to a small number ; (4) it makes a merit of superior natural gifts.

It is a fact that the human intellect has at all times been spurred to its highest exertions by rivalry, contest, and the ambition of being first. The question is, whether a more moderate pitch of excellence, such as befits average faculties, could not be attained without that stimulant. If so, there would be a clear moral gain. Be this as it may, there is no need to bring it forward prematurely, or to press its application at the beginning. In the infant stage, where the endeavour is to draw out the amicable sentiments, it is better kept back. For tasks that are easy and interesting, it is unnecessary. The pupils that possess unusual aptitude, should be incited to modesty rather than to assumption.

The greater prizes and distinctions affect only a very small number. Place-capturing, as Bentham phrases it, affects all more or less, although, in the lower end of a class, position is of small consequence. Too often the attainments near the bottom are *nil*. A few contesting eagerly for being first, and the mass phlegmatic, is not a healthy class.

Prizes may be valuable in themselves, and also a token of superiority. Small gifts by parents are useful incitements to lessons ; the school contains prizes for distinction that only a small number can reach. The schoolmaster's means of reward is chiefly confined to approbation, or praise, a great and flexible instrument, yet needing delicate manipulation. Some kinds of merit are so palpable as to be described by numerical marks. Equal, in point of distinctness, is the fact that a thing is right or wrong, in part or in whole ; it is sufficient approbation to pronounce that a question is correctly answered, a passage properly explained. This is the praise that envy cannot assail. Most unsafe are phrases of commendation ; much care is required to make them both discriminating and just. They need to have a palpable basis in facts. Distinguished merit should not always be attended with pæans ; silent recog-

nition is the rule, the exceptions must be such as to ex-
tort admiration from the most jealous.   The controlling
circumstance is the presence of the collective body ;
the teacher is not speaking for himself alone, but directing the sentiments of a multitude, with which he should
never be at variance ; his strictly private judgments
should be privately conveyed.   Bentham's 'Scholar-
Jury Principle,' although not formally recognized in
modern methods, is always tacitly at work.   The opinion
of the school, when at its utmost efficiency, is the united
judgment of the head and the members, the master and
the mass.   Any other state of things is war : although
this too may be unavoidable.

Punishment.

The first and readiest, and ever the best, form of
Punishment is Censure, Reprobation, Dispraise, to which
are applicable all the maxims above laid down for praise.
Definite descriptions of definite failures, without note
or comment, are a power to punish.   When there are
aggravations, such as downright carelessness, a damaging
commentary may be added ; but in using terms of re-
probation, still more strict regard has to be paid to
discrimination and justice.   The degrees of badness
are sometimes numerical—measured by the quantity of
lesson missed, and the repetition of the failure : this very
definiteness literally stated is more cutting than epithets.

Strong terms of reproof should be sparing, in order
to be more effective.   Still more sparing ought to be
tones of anger.   Loss of temper, however excusable, is
really a victory to wrongdoers ; although for the moment

it may strike terror. Unless a man is of fiendish nature throughout, he cannot maintain a consistent course, if he gives way to temper. Indignation under control is a mighty weapon. Yet it is mere impotence to utter threats when the power of execution is known to be wanting. There is nothing worse for authority than to over-vaunt itself; this is the fatal step to the ridiculous.

Punishments must go deeper than words; indeed, the efficacy of blame depends on something else to follow. Bearing in mind what are the evil tendencies to be encountered in school discipline—want of application being the most constant—we may review the different kinds of penalties that have been placed at the disposal of the schoolmaster. The occasional aggravation of disorder and rebellion has also to be encountered, but with an eye to the main requisite.

Simple forms of Disgrace have been invented, in the shape of shameful positions, and humiliating isolation. As appealing to the sense of shame, these are powerful with many, but not with all; their power varies with the view taken of them by the collective body, as well as with individual sensitiveness. They answer for smaller offences, but not for the greatest; they may do to begin with, but they rapidly lose power by repetition. It is a rule in punishment to try slight penalties at first: with the better natures, the mere idea of punishment is enough; severity is entirely unnecessary. It is a coarse and blundering system that knows of nothing but the severe and degrading sorts.

Detention from play, or keeping in after hours, is very galling to the young; and it ought to suffice for even serious offences; especially for riotous and unruly

tendencies, for which it has all the merits of 'characteristicalness.' The excess of activity and aggressiveness
is met by withholding the ordinary legitimate outlets.

Tasks or impositions are the usual punishment of
neglect of lessons, and are also employed for rebelliousness ; the pain lies in the intellectual *ennui*, which is
severe to those that have no liking for books in any
shape. They also possess the irksomeness of confinement and fatigue-drill. They may be superadded to
shame, and the combination is a formidable penalty.

With all these various resources ingeniously plied—
Emulation, Praise, Censure, Forms of Disgrace, Confinement, Impositions—the necessity for Corporal Punishments should be nearly done away with. In any well-
regulated school, where all the motives are carefully
graded, through a long series of increasing privations
and penalties, there should be no cases but are sufficiently met. The presence of pupils that are not amenable to such means is a discord and an anomaly ; and
the direct remedy would consist in removing them to
some place where the lower natures are grouped together.
Inequality of moral tone is as much to be deprecated in
a class as inequality of intellectual advancement. There
should be Reformatories, or special institutions, for those
that cannot be governed like the majority.

Where corporal punishment is kept up, it should be
at the far end of the list of penalties ; its slightest application should be accounted the worst disgrace, and
should be accompanied with stigmatizing forms. It
should be regarded as a deep injury to the person that
inflicts it, and to those that have to witness it—as the
height of shame and infamy. It ought not to be re-

peated with the same pupil: if two or three applications are not enough, removal is the proper course. The misfortune is that in the National Schools the worst and most neglected natures have to be introduced: yet they should not brutalize a whole school. Even when children are habituated to blows at home, it does not follow that these are necessary at school; parents are often unskilful, as well as hampered in their circumstances, and emergencies are pressing; the treatment at school may easily rise above the conduct of the family. In many instances the school will be a welcome haven to the children of troubled homes; and lead to the generous response of good behaviour.

In point of fact, however, the children of wretchedness are not always those that give trouble, nor is it the schools where these are found that are most given to corporal punishments. The schoolmaster's most wayward subjects come often from good families; and they are found in schools of the highest grade. There should be no difficulty in sending away from superior schools all such as could not be disciplined without the degradation of flogging.[1]

[1] Testimonies are adduced from very distinguished men, to the effect that without flogging they would have done nothing. Melanchthon, Johnson, Goldsmith, are all quoted for a sentiment of this kind. We must, however, interpret the fact on a wider basis. There was no intermediate course in those days between spoiling and corporal punishment: he that spared the rod hated the child. Many ways can now be found of spurring young and capable minds to application; and corporal punishment would take an inferior position in the mere point of efficiency.

It is not to be held that corporal punishment, to such extent as is permissible, is the severest form of punishment that may be administered in connexion with the school. For mere pain, a whipping would often be chosen in preference to the intolerable irksomeness of confinement during

The Discipline of Consequences.

The idea of Rousseau that children, instead of being punished, should be left to the natural consequences of their disobedience, has much plausibility, and is taken up at the present day by educationists. Mr. Spencer has dwelt upon it with great emphasis.

One obvious limitation to the principle is, that the results may be too serious to be used for discipline: children have to be protected from the consequences of many of their acts.

What is intended is, to free parents and others from the odium of being the authors of pain, and to throw this upon impersonal agencies, towards which the child can entertain no resentment. But before counting on that result, two things are to be weighed. First, the child may soon be able to see through the device, and to be aware that after all the pain is brought about by virtue of a well-laid scheme for the purpose ; as when the unpunctual child is left behind. Next, the personifying or anthropomorphic tendency being at its greatest in early years, every natural evil is laid at the door of a *person*—known or unknown. The habit of looking at the laws of nature in their crushing application, as cold, passionless, purposeless, is a very late and difficult acquirement, one of the triumphs of science or philosophy : we begin by resenting everything that does us harm, and are but too ready to look round for an actual person to bear the brunt of our wrath.

play or after hours, and of impositions in the way of drill tasks ; while the language of censure may be so cutting as to be far worse than blows. What is maintained is, that these other punishments are not so liable to abuse, nor so brutalizing to all concerned as bodily inflictions.

A further difficulty is the want of foresight and fore-knowledge in children : they are unable to realize consequences when the evil impulse is upon them. This, of course, decreases by time ; and according as the sense of consequences is strengthened, these become more adequate as a check to misconduct. It is then indifferent whether they are natural or ordained.

Among the natural consequences that are relied on as correctives of misbehaviour in the family, are such as these : going with shabby clothes, from having spoilt a new suit ; getting no new toys to replace those that are destroyed. The case of one child having to make reparation to another for things destroyed, is more an example of Bentham's 'characteristical' punishment.

In school, the discipline of consequences comes in under the arrangements for assigning each one's merit on an impersonal plan ; the temper or disposition of the master being nowhere apparent. The regulations being fixed and understood, non-compliance punishes itself.

CHAPTER IV.

TERMS EXPLAINED.

In discussing Education-questions, there occur certain terms and phrases that suspend great issues, and yet are of ambiguous import. Some of these refer to faculties of the mind, as Memory, Judgment, and Imagination, whose scope needs to be clearly comprehended. Of equal importance is it to fix the meanings to be attached to the words—Training, Culture, Discipline—when opposed to what is expressed by Information.

MEMORY, AND ITS CULTIVATION.

Committing to Memory' is a phrase for learning or acquiring those parts of knowledge that are imbibed without apparently exercising the higher faculties called Reason and Judgment. Such are—names, word lists, in grammar, and in language generally. Likewise the events that we have witnessed impress themselves on our memory, by the mere fact of their having excited our attention. Again, a great part of the early education of children consists in acquiring the fixed arrangements of things that make up their habitual environment. Also, the simpler sequences of cause and effect are laid hold of at first by a mere act of memory.

In order that such acquisitions may proceed rapidly, certain conditions must be fulfilled, formerly described as the conditions of Retentiveness or Memory. The providing for these conditions is sometimes spoken of as exercising the Memory, or cultivating the Memory. Now the question is started—Can we by any artifices cultivate or strengthen the Memory, or the power of Retentiveness as a whole? We may acquire knowledge. Granted. Can we strengthen or increase the natural powers of acquisition? It is, no doubt, said with justice that every faculty can be strengthened by exercise; nevertheless, as regards mental power, the effect is by no means simple.

The absolute power of Retentiveness in any individual mind, is a limited quantity. There is no way of extending this limit except by encroaching on some of the other powers of the mind, or else by quickening the mental faculties altogether, at the expense of the bodily functions. An unnatural memory may be produced at the cost of reason, judgment, and imagination, or at the cost of the emotional aptitudes. This is not a desirable result.

The more common form of exalted memory is the memory for a special subject, which grows by devotion to that subject; being a result of the habits of attention that are engendered towards our leading studies. It is by this artificial strain, that an orator commits his speeches to memory with comparative ease. The memory for places is intensified by habitual attention, the consequence of our special avocations; an engineer or an artist remembers places, not by superior general memory, nor even by particular memory, but by the strain and preference of attention, accompanied by neglect of other matters.

Instead, therefore, of speaking of the cultivation of the faculty of Memory, we should simply consider the means of fostering some definite class of acquisitions, according to the established laws of Retentiveness.

JUDGMENT, AND ITS CULTIVATION.

This is a word employed as a contrast to Memory, and as a synonym for Understanding and Reason. A teacher is expected to cultivate in pupils not only Memory, but also Judgment.

The simplest supposable act of Judging is the comparing of two things, as to their differences, or their agreements, or both. If they are objects of sense, as two shades of colour, it is mere sense discrimination, and depends upon the delicacy of the sense of sight, the amount of attention bestowed, and the close juxtaposition of the specimens. The very same conditions favour the discerning of agreements.

When the two things to be compared are complex objects of sense, as two machines, two houses, two trees, two animals, there are more points to be attended to, but the operation is otherwise the same. When the objects are given, partly by sense appearances, and partly by verbal description of experimental properties, as two minerals, the grasp required is still greater, and the precautions are more specific. It takes an effort to view the complex whole in the advantageous attitude for comparison ; that is, by conceiving the properties of each in the same order. This kind of effort is the result of a mental discipline. The comparison of two cases in law, of two theories in science, of the two different ways of expressing the same doctrine, or explaining the same fact, is a species of judgment, and is favoured by orderly placing of the several circumstances and peculiarities.

Still higher is the act of judging something present to the view by a mental standard of comparison, the result of previous knowledge and experience; as when we judge of the propriety or suitability of a work of in- dustry, or of art, or a scheme of policy. Here we have a wide field of comparison; we have to view the col- lateral facts, and· to follow out consequences. There are other and better ways of describing this mental act than calling it a judgment; it is the application of far- reaching knowledge to determine cause and effect. This cannot be cultivated .as a faculty; it is a vast accom- plishment, the result of prolonged experience and studies in a particular department.

Related to the foregoing is the giving of a sound de- cision in conflicting circumstances; the taking all parts of a case into the account, instead of running off upon one or two points. The man that views a problem thoroughly and exhaustively, omitting nothing that belongs to its solution, is called a man of judgment; but there is equal impropriety in styling this a faculty, and in speak- ing of improving it as such.

These are very high instances of the Judging faculty, being nothing short of the utmost maturity of the human understanding, in particular departments of affairs. It is out of the province of the schoolmaster to speak of this kind of judgment. There is a more familiar, but loose application of the word, brought out by the contrast with

Memory ; the power of comprehending as opposed to the power of remembering by rote. This is a real and important distinction, better expressed by Understanding, or Comprehension, than by judgment. It often occurs to us in teaching to have to test a pupil's understanding of a passage, a principle, or a rule, that has been committed to memory.

Reason, *Reasoning*, and *giving Reasons* are intellectual operations not far removed from some of the meanings of Judgment. . They are much more definite and precise, in consequence of their use in Logic ; by reference to which they can be learned, better than in any other way.

IMAGINATION.

A wide word. It covers operations very various in kind, and its employment is calculated to obscure some of the most critical processes in education. It has in some of its meanings a very lofty function ; in others, it expresses the utter degradation of the human powers.

'Without imagination,' says Godwin, 'there can be no genuine ardour in any pursuit, or for any acquisition, and without imagination there can be no genuine morality, no profound feeling of other men's sorrows. no ardent and persevering anxiety for their interests.' This definition trespasses upon a number of distinct mental powers, including all that is comprised in the name ' Sympathy.'

The first meaning of Imagination is expressed also by Conception, or the Conceiving Faculty, whereby we realize a picture of what we have not seen ; the usual medium of presentation being language, with or without the aid of pictorial sketches. This is a power that grows with our experience of scenes and situations, and depends on the goodness of the pictorial memory. Excepting by increase of knowledge, there is scarcely any existing provision for cultivating or for augmenting the power; the schoolmaster could do little for it, if he were to try. A learner may be exercised in conceiving things from descriptions; but the real art involved in the case is the art of description itself.

To enter into or conceive other people's feelings is an exercise in Sympathy or Moral Education, and is also a means to our enjoyment of history, poetry, and romance. It is one of the consequences of our life experience, our social dispositions, and our acquired knowledge, but is not easily brought under school lessons. Like moral teaching generally, it may be quickened by an apt teacher in some happy moment, but cannot, so far as I am aware, be made to recur upon a pre-arranged plan.

There is the same error in speaking of Conception, as a faculty to be trained, that was already remarked upon under Memory. The pupil may be assisted to conceive certain things, as a ship of war, a tropical forest, or a paradise; the instrumentality being language well employed, together with sketches or pictorial designs. But this is not cultivating a faculty of conceiving, further than that one successful effort facilitates others of a like kind. As a systematic practice, it is not embodied in any form of teaching at present in use: although incidental to many of our school exercises, it is not steadily followed out in any.

The highest meaning of Imagination is the creative

faculty of the poet or the artist, which passes entirely out of the reach of express training, although all the ways of storing the mind may contribute to it. It is excluded from schemes of Education as too high for the school. But it contains an element in common with the lowest and easiest efforts of Imagination—the element of Feeling, or Emotion, which distinguishes Art constructions from the creations of the man of science. Every work of art gratifies some of our strong emotions—Love, Anger, Revenge, Sublimity, the Ridiculous, and many others. A scientific invention is a matter of mere utility, as measured by pounds, shillings, and pence.

In the following out of this emotional gratification, the imaginative artist takes great liberties with his subject ; he exaggerates, re-arranges, admits fiction and extravagance, and is nowise bounded by reality or truth. This is what makes the charm of imaginative literature, to the young especially. Far more powerful emotions may be roused by romance than by anything in the actual world of events ; and it is these higher effects that we covet. This, however, is not so properly cultivating Imagination as revelling in it ; extracting emotional excitement by means of it. Indulging imagination is indulging the emotions, and the only question is—which emotions ?

This kind of Imagination is to be viewed, in the first instance, as a source of pleasure, an ingredient in the satisfaction of life. In addition to our enjoyments gained from contact with realities, we crave for the contribution that comes from ideality. Now, ideality is a different thing for different ages: fairy-tales and extravagances for the young ; the poetry of Milton for the old. There is nothing educative, in the first instance ; we are not aiming at instruction, but drinking in emotion. The gratifying of children with the literature of imagination is a matter for the parent, as much as giving them country walks or holiday treats. It has a good side and a defective side ; the balance is struck, in any adequate discussion of the uses and abuses of 'Fiction' as a whole.

While the basis of imaginative interest is the stimulus given to emotion, there is a certain intellectual element, in the pictures, scenes, and incidents that strike out emotional sparks. These are impressed on the memory through the excitement that they cause ; they become a part of the intellectual furniture, and may be afterwards turned to account. They may assist in imaginative creations of our own, and be used as illustrating and adorning the sober truths of reason. When we come to fictions of a lofty order, as the works of the great poets, we store up still more exquisite pictorial combinations ; we imbibe into our recollections the highest strokes of human genius. Here, then, fiction is an element in our education. How far it should be given in the school is a question for future consideration.

It is apparent, however, that the conceptive faculty, previously spoken of, is aided by the presence of strong emotion ; but if aided, it is also limited and biassed. The conception is taken in, just as it suits the feeling, and no further. The conception of a battle-scene is a great effort of mental constructiveness ; but it is never complete. The portions that excite the most intense interest are conceived with some tolerable vividness ; we pick and choose the most sensational turns of the action ;

but of the general arrangements we carry away only the feeblest notion.

The intervention of the schoolmaster in the culture of imagination ought to repress the extravagant emotional preferences, and favour the complete and impartial exercise of that really great function of intellect—the power of conceiving, in all the exact lineaments and proportions, scenes and events that have not been experienced—the historical imagination, as distinguished from the poetical. Without rejecting the aid of emotional interest, an instructor endeavours to counterwork its bias and partiality, not to speak of its distortion and falsification of reality. The power of full concrete realization is a high effort of mind, rarely attained even by the educated: it is a talent in itself; and the snatches of fairyland engrained by the emotions of the marvellous are but the faintest approaches to such a power.

PROCEEDING FROM THE KNOWN TO THE UNKNOWN.

This is a favourite maxim of the teaching art, but it is seldom set forth in a way to afford definite guidance. There is a plain enough meaning in easy cases : an explanation should consist of references to facts already understood, otherwise it cannot itself be understood. This is merely the law of progress from the elementary to the composite ; to master one stage of advancement before proceeding to the next. Anyone that consciously violates such a plain requirement is hopeless ; and he that could lay his hand on his heart and say that he never once violated it, would merit immortal remembrance.

If a demonstration proceeds upon principles not already understood ; if a description contains terms with
no meaning to the person addressed ; if directions involve acts that have not been previously performed, the
upshot is a failure. In the stage where instruction is
given in the strict methodical form, as in a regular
course of science, the consecutive arrangement is more
or less adhered to, yet often not without difficulty. In
the previous stages, where knowledge is given by
snatches, there is no security for the right order. In fact,
the most immature minds are exposed to the greatest
jumbles ; and one may wonder how they can imbibe
information so supplied. The course may be inevitable,
but it is not advantageous ; and it will be necessary, in
a subsequent chapter, to consider fully why the necessity
arises, and how its bad effects may be mitigated.

In speech we can but bring forward one thing at a
time ; our facts and statements follow a serial order.
Now, for the comprehension of a difficult subject, it
would sometimes be desirable that two or three things
should come abreast, and be simultaneously conceived.
This is one obstacle that the pupil has to overcome.
Another evil is, that our clearest and most elementary
statement may unavoidably introduce matter that has
not previously been comprehended, and so leave a blur or
dark spot, obscuring all that follows, until such time as
we come across something that lightens the darkness.

ANALYSIS AND SYNTHESIS.

These words are freely made use of in laying down
directions for the guidance of the teacher. The meanings usually assigned to them are very hazy. ' Analysis

is the more distinct of the two; there being certain specific and well-known examples, as, 'the Analysis of the Sentence,' in grammar. Analysis applied to reading-lessons is not quite so obvious; the meaning suggested is that a complex object is viewed in its separate parts. Thus, a steam-engine could be analyzed into the cylinder, the parallel motion, the fly-wheel, the governor balls, and so on. This, however, does not need such a high-sounding name. 'Description' would suit quite as well.

The most scientific meaning of Analysis is that connected with the process of Abstraction. A concrete substance, as a mineral, or a plant, is analyzed into its constituent properties, by successive abstractions, whereby each in its turn is viewed apart. Such is the Natural History description of a mineral—the enumeration of its properties, mathematical, physical, and chemical. For this use the word is somewhat superfluous, and therefore misleading.

There is another form of Analysis, in the separation of a complex effect into its constituent effects, as in the revolution of the motion of a planet into the two tendencies centripetal and centrifugal. In the same sense, we may speak of analyzing a man's character or motives. So we may analyze a political situation, by assigning all the influences at work.

These meanings are all distinct enough, but they are not all in want of this special name, being otherwise provided for. Except in the well-understood instance of Grammatical Analysis, and in the last-mentioned case of composition of forces or agents, we should be much better without the word. Chemical Analysis and

Geometrical Analysis, are peculiar cases that need not be discussed.

More trouble is given by the word ' Synthesis,' although it ought to be in all respects the opposite or obverse of Analysis. There is a Grammatical Synthesis, worked out by Mr. Dalgleish into an exercise in grammar, in which the members of a sentence in separation are to be put in their places again. In the resolving of a complex object into its parts, with a view to orderly delineation, as a machine, there is no corresponding synthesis : the word has no meaning.

The abstractive separation of properties does not need any synthesis. When abstraction prepares us for making an inductive generalization, like the Law of Gravity, there is a counter process of *deductive* carrying out of the law to new cases, and this may be called synthesis; but 'deduction' is a better word.

When **we** analyze the forces at work in an operation —physical, mental, or social—we do not need to compound them again, unless we were supposing new situations, where the composition is differently made up. We could chalk out the orbit of a planet whose solar distance and other elements were different from the case of any known planet.

To express the conduct of any school-lesson, under either of the terms Analysis and Synthesis, is to produce the utmost confusion in the mind of a young teacher : as everything that the words cover is conveyed by other names, more expressive and more intelligible; such are Description, Explanation, Abstraction, Induction, Deduction.

OBJECT LESSONS.

Considerable ambiguity attaches to the phrase ' Object Teaching.'   It seems to have come into use through
Pestalozzi's system of imparting the abstractions of
Number, &c., by concrete examples.   This is a perfectly intelligible meaning, and lies also at the foundation of all teaching of general knowledge.   In the
carrying out of the system, the teacher brings forward
such a selection of concrete objects as concur in some
one general impression, notwithstanding great differences
in other respects.   To impress the number ' four,' a
great many groups of four would be presented to the
pupil ; to impress the notion of a circle, many round
objects would be adduced, differing in size, material,
and other points.

The Object Lesson represents a totally different line
of tuition, when it is looked upon as cultivating the
senses or maturing the observing faculties.   The former
case related to generalities, this refers to specialities
When a pupil is set to discriminate nice shades of colour,
or differences of musical tone, it is by presenting these to
the senses and inciting the attention upon them.   How
far this is requisite in ordinary school education, is a
doubtful matter.   When a special art is taught, as
painting or music, the delicate discrimination of degrees
of colour or tones is a part of the teaching; but with a
view to knowledge of the world, the same special training may not be necessary, except on set occasions or for
select purposes.   It is no part of the highest knowledge
of things in general to possess a delicate tact in mea-
suring lengths by the eye or weights by the hand.

At all events, this special acquirement does not need the designation ' Object Lessons ' to set it forth. The 'cultivation of the senses' is a more suitable mode of describing it ; and such cultivation is a very intelligible form of training, if we only satisfy ourselves that it is required.

A third aspect of the Object Lesson has reference to the acquisition of Language, which, in the first instance, is the associating of things with their names. In order to connect a word with a thing, we must have some notion of the thing ; by sense, observation, or in whatever way it comes before us. The names first learnt are the names of familiar surrounding objects, for the most part individual and concrete. Attention is directed upon the objects, and the names are pronounced, and there is a speedy union of the two in an act of association or memory. To extend the knowledge of language is to extend the knowledge of objects ; and, according as we have opportunity of bringing forward new objects and of getting them attended to, we enlarge the intelligent use of language, with which goes hand in hand a knowledge of the world, at least so far as concerns the characteristic properties of things. To use names properly we must not confound things that differ ; we must know enough for discrimination, although we may not know everything. The dog must not be mistaken for the cat, nor the lamp confounded with the fire.

It does not seem, at first sight, as if the lessons of the schoolmaster could contribute much to this species of object knowledge. It grows out of the whole experience of the waking life.

Moreover, Object Knowledge is scarcely the proper

designation for this case either. Many of the things noted, and especially the earliest, are the individual things about us ; but to mark these is only a first stage, and is followed up by a grander operation. General names soon come to be used and understood by every child. No doubt these correspond, at first, to very easy gene-ralities—light, dark, big, little, chair, spoon, doll, man, water, and so on ; nevertheless, they have to be under-stood, not by looking at one thing, but by comparing different things, attending to agreements and abstract-ing differences.

Again, it has been pointed out, that many terms express *situations* of things rather than things them-selves. Such are all the names for time, and place, and circumstances. Yesterday, to-morrow, are not objects ; out of doors, in the house, are situations. Actions, in like manner, involve a distinct mode of looking at the world ; 'drink,' 'stand,' 'come,' 'speak,' 'cry,' 'bring,' are intelligible to the youngest child, from observing a whole configuration or group of appearances.

Once more, what shall we say to *subject* states, which come early to the foreground, along with the things of the outer world? The elementary states of being pleased and pained, of liking and disliking, are noted by the infant experience, and come to be matters of mutual understanding at a very early stage.

Thus, in the learning of language, there are various well-marked operations that ought to be viewed by themselves, as they enter into the economy of teaching, and that ought to be described in appropriate terms, and not by the very ambiguous and misleading phrase —Object Lessons.

Another meaning remains. I allude to the practice of giving exhaustive lessons, on selected objects, in the course of the ordinary teaching. Thus, to take a piece of Coal. A specimen is produced, and attention is called to its appearance and various sensible properties; by which the pupil is, perhaps, made to inspect it more narrowly and curiously than before. Thus far, we might call it a Sense, or Observation, lesson. But the master does not stop here: he goes minutely into the Natural History and Chemistry of Coal; describing where it comes from, how it is first produced, what uses it serves, and by virtue of what properties it serves these uses. This may be called an Information lesson on an object. A chosen substance is made the text for a dissertation on natural processes and properties that implicate many other substances. We cannot give the history of coal without adverting to the Structure of Plants; nor can we assign its uses without bringing in Chemical Union and Heat. The propriety of this kind of lesson depends on various circumstances; and the consideration of these will come up again. It is not so much an object lesson, as the employment of an Object Text. It enables us to group a mass of very abstract notions and properties round a concrete unit. It is a convenient form of the popular lecture, as when we see Professor Huxley or Dr. Carpenter adopting, as a topic of discourse, a piece of Chalk.

INFORMATION AND TRAINING.

The contrast between Information, on the one hand, and Training, Discipline, calling out the Powers or Faculties, on the other, is one that plays a great part

11

in discussions relative to education. It is expedient to clear up this distinction in the only way that it can be done, namely, by examples of the two effects. The allegation is often made, that there are kinds of instruction of little value as information, but of such supreme efficacy as discipline, that they are to be preferred to any species of information that is nothing beyond information. The object of education, it is said, is not to instil truths, but to draw out and exercise the mind's faculties and forces.

First, as to Information. The Arithmetical operations of adding, multiplying, &c., taught for the purposes of practice, without any reference to their foundations or principles, would probably be regarded as useful information, but not as discipline. This is, in fact, the way that they are apprehended by the great majority of pupils.

Again, in regard to our own language. All the usages of the language, including the highest rules of correctness and propriety, may be imparted merely as guidance in speaking and writing with exactness; there being no attempt to cast them into a methodical shape or to reduce them under rational explanations. This would be pure information; the teaching of language, so conducted, would be very useful, but would not be called a mental discipline. Those persons that all their life have been associated with only such as speak correctly and elegantly, become correct and good speakers without any training at all. A foreign language might be imparted in the same way; even the dead languages could be taught without grammars or rules; that is to say, by mere habituation in reading books.

Historical facts are for the most part nothing more than information. They must follow the order of time, but this constraint does not amount to a mental discipline. Chronology informs us of the sequence of the greater events of History; it is laid up in the memory as so much information, and does not aspire to draw forth or cultivate any faculty; it is merely one of many ways of calling memory into use.

Geographical facts may be simply matter of information. In so far as they are devoid of connection and system, they are information solely. In so far as they can be embodied in an orderly scheme—a descriptive method, which facilitates both the recollection and the understanding of them, they rise to some sort of training. They require the pupils to master the scheme, and so give them possession of it, as an art that they may themselves employ in dealing with similar details.

All those facts that relate to useful operations in the arts of life, that serve to guide artificers in their work, and to instruct everyone how to attain desirable ends, constitute a vast body of useful information. The recipes of cookery, the arts of husbandry and of manufacture, the cures of disease, the procedure in courts of law, are most valuable as information; but they are not regarded as giving us any form of discipline. Our books of household management, of horticulture, and of the rearing of animals, are full of the best possible information, but nothing more.

Even in the Sciences, properly so called, there may be matter that a pupil imbibes merely as so much information. The practical conclusions from scientific principles may be seized and turned to account, while the demonstrations, deductions, and proofs are wholly missed, as already mentioned in regard to Arithmetic. Even in geometry a student may carry away with him the theorems, as so many truths applicable to practice, without understanding their dependence upon each other; in other words, without knowing geometry as a science. So we may have a large stock of physical, chemical, and physiological facts, and may be quite correct in our statements of them, and yet may not know any one of these sciences. The same remark applies to the knowledge of mind.

Nevertheless, it is not a low order of intelligence that has taken in, remembered, and is able to apply an extensive stock of maxims of practice and utility in various departments. There may not be anything amounting to high discipline, but there is an expenditure of good intellectual force. The higher the character of the work, the more scope is there for fine discrimination or accurate perception, in order to suit the means to the end. Navigating a ship, practising physic, may be on the basis of information alone; but it is a superior order of information. There is, in short, a scale of amount and difficulty, in regard to what we may consider as mere information; and when we touch the higher degrees, we come upon something that involves the best faculties or forces of the mind.

The truth is, that for the higher professions the extent of practical knowledge is such that it cannot be comprehended, held together, or rendered sufficiently precise, unless we have a certain amount of science and scientific method, such as would probably come within the scope of Discipline.

Let us now review the meanings of DISCIPLINE. While the mere facts of science, turned to account in practical operations, are called information, the *method* of science, the systematic construction of it, the power of concatenating and deriving truths from other truths, is treated as something distinct and superior.   The thorough comprehension of the method of Euclid, the tracing of the Arithmetical and Algebraical rules to first principles, would be considered as training, discipline, the calling forth of the powers.

Most definitions of training are obscured through the mode of describing mind by faculties.   We have seen that to train 'Memory' is a very vague way of speaking.   Equally vague is it to talk of training Reason, Conception, Imagination, and so forth.   Moral training is much more intelligible; there is here a habit of suppressing certain active tendencies of the mind, and fostering others; and this is done by a special discipline —like training horses or making soldiers.   The analogy is not very close between these exercises and the improvement of the intellectual powers; still, such as it is, it is illustrative.   To train a soldier is to bring him to the ready performance of a number of combined movements, to which he is led on by graduated exercises. Head knowledge, or information, is combined with the training, but is a distinguishable factor.   In all other skilled avocations, a similar element of training is present.   In many, however, the muscular aptitudes do not form the main part; the training penetrates more into the thoughts or ideas.   For example, the training of an officer is more mental than bodily; it consists in a knowledge of the configurations, movements, groupings of bodies of men; and a readiness to direct the proper movement in the proper situation. It is knowledge realizable in practice with the quickness of an instinct.

While science, as already noted, may be imbibed in a form that does not pass beyond information, the arts of scientific observation and research imply training proper. The senses have to be exalted, the attention directed, methods of procedure learnt, to the pitch of habituation; with all which there concurs much information of details, but the information is distinct from the training.

In the vast accomplishment of Speech, we can enumerate various things properly designated training. Elocution, or the management of the voice, would be considered as training throughout. The knowledge of detached names would exemplify information pure and simple; it is bare word memory. The arrangement of words in sentences, with attention to grammatical forms and all the other proprieties of speech, would be accounted training. The still higher arts of arranging the thoughts in lucid expression, if known only as rules or theory, might be called information; but when embodied in habits would rank as training. Hence we speak of a trained orator or writer. It is in this sense that moral teaching and moral training are totally distinct things.

The element of Form, Method, Order, Organization, as contrasted with the subject-matter viewed without reference to form, has a value of its own; and any material that displays it to advantage, and enables it to be acquired, is justified by that circumstance alone.

The targets used in learning to shoot, the wooden soldiers that are aimed at in sabre drill, although unreal,
are effectual.

Worth belongs to any subject of study if it conveys
methods that are useful far beyond itself. The sciences
that embody an organization for aiding the mind—
whether in deductive method, such as Geometry and
Physico-mathematical Science; in observation and induction, as the Physical Sciences; or in classification,
as the Natural History Sciences;—would on these
grounds alone be admitted to the higher circle of
mental Discipline or Training, irrespective of the value
of the facts and principles viewed separately or in detail.
It depends partly on the teacher and partly on the
scholar whether the element of method shall stand
forth and extend itself, or whether the subjects shall
only yield their own quantum of matter or information.

Logic is nothing, if not training. The information
mixed up with it is all to be used for training purposes.
It is the element of scientific *form*, which is more
thoroughly impressed by being singled out for special
consideration. It is the grammar of knowledge.

There is a form of mental efficiency that attaches
more or less to every productive effort—the giving attention to all the rules and conditions necessary for the
result intended. We cannot perform any piece of work
unless we are alive to everything that is involved in it ;
we cannot guide a boat, unless we manage sail and
rudder according to the direction of the wind. We
cannot turn out a good sentence without fulfilling numerous conditions. When we follow written rules, we must
interpret them correctly, and apply them appositely

This is a discipline that we learn from everything that we have to do; it is not a prerogative of any one study or occupation, and it does not necessarily extend itself beyond the special subject. Because a man can hunt well, it does not follow that he shall be a good politician or a good judge; although in all these functions there is the common circumstance of taking account of every condition that enters into a given effect. A very superior mind, like Cromwell's, probably transfers the conditions of efficiency from one department to others remote from it; and thus becomes rapidly developed into fitness for new domains of practice.

In our subsequent review of Education Values, the difference between Information and Training will be rendered still more precise.

ONE THING WELL.

This is a favourite commonplace of teaching and of private study. It proceeds upon the idea that it is better to be thoroughly versed in one limited walk of knowledge or culture, than to pass slightly over a wider area.

There are different senses attached to learning anything well. In the first place, it may mean simply that full habituation to any piece of knowledge or practice that makes it a matter of mechanical certainty and ease; as in the case of a thorough proficient in Arithmetical sums. Long iteration has this effect in everything; and it is indispensable in matters of business occupation.

In the second place, there is a higher form of learning anything well, by which is meant a full and minute acquaintance with all details, qualifications, exceptions.

and whatever is included in the complete mastery of a large and complicated system. This is the equipment of a thorough lawyer or a thorough physician, who must know both leading doctrines and their varying applications to a wide host of varying circumstances. So much is involved in these professions, that no man is expected to be versed in more than one. So, in any of the leading sciences, there is a kind of mastery that is equally multifarious and absorbing; whether it be the innumerable combinations of mathematical formulæ, the vast details accompanying an experimental science, like Chemistry, or the seemingly still more inexhaustible fields of Botany and Zoology. To be thoroughly accomplished in any one of these branches, we must be content with a limited acquaintance with the other departments. The expression for this higher knowledge is more properly *multa* than *multum*; the field may be circumscribed, but its minute and exhaustive survey implies a multitudinous knowledge. For tne highest uses of a science, this is the only knowledge that avails.

In the point of view of information, a single fact, if clearly understood, is of value, although no other be drawn from the same source. And even as regards discipline, in the acceptation of special method, this is best learned in a select and limited portion of material; as when we study classification from Botany. In neither of these respects is it necessary to spend time over an exclusive subject. Having a definite purpose, we must pick and choose at many points, and the present maxim is without relevance.

In a right view of scientific education, the first principles and leading examples, with select details, of all

the great sciences, are the proper basis of the complete
and exhaustive study of any single science. This may
not be apparent in Mathematics, the first of the fundamental sciences, but it applies to all beyond. A man
cannot be a good chemist without possessing, on the
one side, a fair knowledge of Physics, grounded in Mathematics, and, on the other, some acquaintance with
Physiology. The thorough knowledge of every subject
implicates everything that leads up to it, as well as
everything that can throw side-lights upon it; although,
of course, these aiding subjec s are not mastered to the
same extent as the subject that they are intended to
assist. In almost all departments of study there are
gradations of acquirement, each thorough and sufficient
for a given purpose. This is least true of languages;
seeing that, till we have reached the point when a
language can be used in communication, we have done
little or nothing.

In the situation of the beginner in any branch of
knowledge, it is expedient to abide by one course, one
scheme, one book, although not absolutely perfect. When
the very groundwork has to be laid, distracting views are
to be avoided. Before criticizing, controverting, or amending a system, the teacher should make his pupils perfectly
familiar with its details. In Geometry, Euclid, or whatever other book is chosen, is verbally adhered to, as if
it were an infallible revelation; when once thoroughly
known, defects may be pointed out, and alternative lines
of demonstration indicated. It is very desirable, notwithstanding, that the book so used should have as
few defects as possible. The principle contended for
by De Morgan, that Euclid is the best book for begin-

ners because of its defects, is not merely paradoxical but positively unsound. It proceeds upon the admitted necessity of finding some exercises for the pupils' own powers, in teaching a science. But such exercises can be obtained apart from the blunders of a text-book; which blunders, being unintended, cannot, except by the barest accident, answer any purpose of tuition.

The present doctrine is abused in the great English schools by being made the pretext of narrowing the studies to the old traditional classics, as against the admixture of science and modern thought. The allegation is that two or three subjects well taught—meaning Latin, Greek, and Mathematics—do more good than six or seven less well taught, although these may include English, Physics, and Chemistry. The same narrowing tendency is repeated, from the modern side, in the demand for very minute and practical knowledge in such vast and absorbing subjects as Chemistry, Physiology, Zoology. In estimating the value of a branch of study, we must consider not merely what it gives us, but what, through engrossment of our time, it deprives us of.

The *multum non multa* is in curious conflict with the most popular current definition of education—the harmonious and balanced cultivation of all the faculties.[1]

---

[1] There is much force in the following observations quoted from Mr. T. Davison, in Mark Pattison's article on Oxford Studies in *Oxford Essays*, 1855: 'A man who has been trained to think upon one subject, or for one subject only, will never be a good judge in that one; whereas the enlargement of his circle gives him increased knowledge and power in a rapidly increasing ratio. So much do ideas act, not as solitary units, but by grouping and combination; and so clearly do all the things that fall within the proper province of the same faculty of the mind, intertwine with and support each other. Judgment lives as it were by comparison and discrimination.'

CHAPTER V.

EDUCATION VALUES.

I WILL now glance at the leading branches of human culture, with a view to seize the characteristic mental efficacy of each. I do not propose to take up every assignable acquisition, but merely those things that enter into the ordinary course of school education. There are various departments of valuable training that properly come under individual self-culture; such are games, arts, and accomplishments.

The carrying out of our design involves a full consideration of the two leading departments, SCIENCE and LANGUAGE. These comprise the great mass of human information in its purest types, and should be thoroughly appraised before entering upon mixed subjects, such as . Geography and History. Fine Art will be touched on, and then adjourned to a chapter apart. The mere purely mechanical acquirements, as Drawing and Handicraft, will be considered only in their subservience to Intellectual Education.

THE SCIENCES.

Of Science generally we can remark, first, that it is the most perfect embodiment of Truth, and of the ways of getting at Truth. More than anything else does it impress the mind with the nature of Evidence, with the labour and precautions necessary to prove a thing. It is the grand corrective of the laxness of the natural man in receiving unaccredited facts and conclusions. It exemplifies the devices for establishing a fact, or a law, under every variety of circumstances; it saps the credit of everything that is affirmed without being properly attested.

Before the birth of science, and in minds debarred from scientific training, the greatest security for truth has been practice. We cannot secure any practical end in this world without observing the natural conditions ; we must estimate the force of a current in order to build a rampart that will stand ; we must know the motives of a man before we can secure his services. In proportion to our regard for truth, and to our means of ascertaining what is true, is our power over the material and the moral world. The greatest test of our knowledge is the test of practical fulfilment ; this is the scientific man's test ; so that the man of practice and the man of science have this much in common.

The defect of the practical man is the limitation of his tests to his own sphere of working; he seldom learns to extend his methods into other spheres. It is possible to be a good engineer and, at the same time, a very prejudiced judge of the human feelings. An accomplished lawyer is not necessarily a good administrator.

The second great liberalizing feature of Science is its mode of setting forth general or generalized knowledge ; the antithesis of the individual and the general, with the gradations of generality and the various relations of co·

ordination and subordination, constituting the heart and soul of method in grappling with multitudinous and complicated facts. The untrained mind confounds general and particular, co-ordinate and subordinate, in one inextricable jumble. It is through science that we take the best grasp of the method of unfolding a subject from the simple to the complex.

In reviewing the Sciences in order, we may divide those relating to the outer world under three groups : Mathematics, as representing *Abstract* and *Demonstrative* Science ; the *Experimental* Sciences—Physics, Chemistry, and Physiology ; and the Sciences of *Classification*, commonly called Natural History. The Science of Mind will be taken apart.

The Abstract Sciences.

MATHEMATICS, including not merely Arithmetic, Algebra, Geometry, and the higher Calculus, but also the applied Mathematics of Natural Philosophy, has a marked and peculiar method or character ; it is by pre-eminence *deductive* or *demonstrative*, and exhibits in a nearly perfect form all the machinery belonging to this mode of obtaining truth. Laying down a very small number of first principles, either self-evident or requiring very little effort to prove them, it evolves a vast number of deductive truths and applications, by a procedure in the highest degree mathematical and systematic. Now, although it is chiefly in the one domain of Quantity, that this machinery has its fullest scope, yet, as in every subject that the mind has to discuss, there is a frequent resort to the deductive, demonstrative, or downward

procedure, as contrasted with the direct appeal to observation, fact, or induction, a mathematical training is a
fitting equipment for the exercise of this function. The
rigid definition of all leading terms and notions ; the
explicit statement of all the first principles ; the onward
march by successive deductions, each one reposing on
ground already secured ; no begging of either premisses
or conclusions ; no surreptitious admissions ; no shifting
of ground ; no vacillation in the meanings of terms,—all
this is implied in the perfect type of a deductive science.
The pupil should be made to feel that he has accepted
nothing without a clear and demonstrative reason ; to
the entire exclusion of authority, tradition, prejudice, or
self-interest.

Such is, to a very considerable degree, the impression
made by a course of mathematical instruction. It would
be made in a still higher degree if the science were more
true to itself, and did not permit a certain looseness
of treatment in the Definitions, and still more in the
Axioms ; while, in the demonstrations, merely verbal
transitions are sometimes given as steps of demonstrations. These deficiencies will, in time, be remedied, and
the science will then be, what it scarcely is yet, an embodiment of pure Deduction.

In addition to this general view of Demonstrative
Reasoning, the details of Mathematical science contribute some of the most valuable materials towards the
building up of the reasoning powers.

For example, it is here that we begin to understand
the manner of handling concurrent elements. We have
a result determined by two or three factors, and we
learn to compute the bearing of any change in any one

or more of these factors. We find one or two of them remaining unchanged, and yet the result varies because of a change in the third; we see how all may change and the result remain constant, from the changes being such as to neutralize each other, and so on. The steady application of this simple process to the more complicated operations of nature and of mind distinguishes the educated intellect. The exercise is still further carried out in Mechanics, in connection with forces, and is thus made still more pointed in its ulterior applications.[1]

[1] Take the following as further instances. The heat of any given day is due partly to the position of the sun, corresponding to that day, partly to atmospheric causes, of which the chief is the prevailing wind. 'When a course is set in motion, and acts constantly in one direction with a steady and uniform nisus, its operation may sometimes be suppressed by an overwhelming opposition, sometimes repressed and weakened, though not quite overborne, by impeding and retarding forces. Thus the fear of punishment is a cause constantly acting in the same direction. Its tendency is always to deter; but this tendency may be counteracted in each actual case by a variety of circumstances which sometimes weaken and sometimes nullify its operation. Freedom of trade tends constantly to facilitate supply, and thus to produce cheapness; but in considering the probability of this effect being produced in any individual case, we must estimate the probability of such events as deficient harvests, difficulty of freight, insecurity arising from war or civil commotions, and the like. The reduction of a high customs duty on any article would naturally tend to increase its importation; yet a change in the public taste, or the discovery of a cheaper or preferable substitute, might prove an effectual counteraction to this tendency.'—*Lewis's* '*Methods of Observation and Reasoning in Politics*,' vol. ii., p. 171.

The phrase 'cæteris paribus' (other things remaining the same) is a mathematical coinage, for guarding against the error of supposing that a course will produce its effect under all circumstances indiscriminately.

In Addison's *Essays on the Pleasures of the Imagination* there is a triple conversation, which he manages thus :—'I shall first consider those pleasures of the imagination which arise from the actual view and survey of outward objects; and these, I think, all proceed from the sight of what

What renders a problem definite, and what leaves it indefinite, may be best understood from Mathematics. The very important idea of solving a problem within

is great, uncommon, or beautiful. There may, indeed, be something, so terrible or offensive, that the horror or loathsomeness of an object may overbear the pleasure which results from its greatness, novelty, or beauty ; but still there will be such a mixture of delight in the very disgust it gives us, as any of these three qualifications are most conspicuous and prevailing.' This is a slight and passing testimony to the principle of composite forces. A mind trained on Pure and Applied Mathematics would make it play a leading part all through the investigation.

Take, as another example, the complex influences that enter into Nationality, as expressed by J. S. Mill:—'A portion of mankind may be said to constitute a Nationality, if they are united among themselves by common sympathies, which do not exist between them and any others— which make them co-operate with each other more willingly than with other people, desire to be under the same government, and desire that it should be government by themselves, or a portion of themselves, exclu- sively. This feeling of nationality may have been generated by *various causes.* Sometimes it is the effect of identity of race and descent. Com- munity of language, and community of religion. greatly contribute to it. Geographical limits are one of its causes. But the strongest of all is identity of political antecedents ; the possession of a national history, and consequent community of recollections ; collective pride and humiliation, pleasure and regret, connected with the same incidents in the past. None of these circumstances, however, are either indispensable, or necessarily sufficient by themselves.' For handling a discussion of this nature, a knowledge of the facts is not enough ; there must be also a firm grasp of the conception of concurring elements, and of all the varying results that different proportions of these may bring about. The groundwork of this conception is best and soonest got at by a discipline in the Mathematical Sciences.

In laying down the indications of depravity in men's dispositions, Bentham uses, not improperly, the forms of mathematics. For example : ' The strength of the temptation being given, the mischievousness of the disposition manifested by the enterprize, is as the apparent mischievousness of the act.' Or again : ' The apparent mischievousness of the act being given, a man's disposition is the more depraved, the slighter the tempta- tion by which he has been overcome.'

Aristotle, in Book I. chap. vii. of the ' Rhetoric,' compares different degrees of what is good, and makes great use of mathematical forms.

12

limits of error is an element of rational culture, coming from the same source. The art of totalizing fluctuations by curves is capable of being carried, in conception, far beyond the mathematical domain, where it is first learnt. The distinction between laws and co-efficients applies in every department of causation. The theory of Probable Evidence is the mathematical contribution to Logic, and is of paramount importance. The remark of Gibbon was very loose, that mathematics disqualifies the mind for dealing with subjects where we attain only pro-bability.

All this supposes Mathematics in its aspect of *train-ing* ; or as providing forms, methods, and ideas that enter into the whole mechanism of reasoning, wherever that takes a scientific shape. As culture imposed upon everyone, this is its highest justification. But, if so, these fruitful ideas should be made prominent in the teaching ; that is to say, the teacher should be fully conscious of their all-penetrating influence. Moreover, he should keep in view the fact that nine-tenths of pupils derive their chief benefit from these ideas and forms of thinking, which they can transfer to other regions of knowledge ; for the large majority, the solu-tion of problems is not the highest end.

In the point of view of information, the uses of Mathematics are more obvious ; but these uses, when carried to their utmost stretch, suppose special profes-sions. For people generally, facility in Arithmetical sums is very desirable, and this is greatly increased when the study is prolonged into the higher branches of Algebra. Geometry, besides its uses to the land surveyor, engineer, navigator, and many others, has a more general utility in enabling one readily to estimate forms, distances, situations, and configurations, whether on the small scale or the great. In the examples of Arithmetical and Algebraical operations, much valuable practical knowledge is incidentally obtained; and more might be done to turn the opportunity to account.

Those that can readily master the difficulties of Mathematics find a considerable charm in the study, sometimes amounting to fascination. This is far from universal; but the subject contains elements of strong interest of the kind that constitutes the pleasures of knowledge. The marvellous devices for solving problems elate the mind with the feeling of intellectual power; and the innumerable constructions of the science leave us lost in wonder.

There are advantages claimed for Mathematics that are not specially confined to it. For example, in following out a long demonstration, the power of sustained attention is necessary; but there are fifty things besides Mathematics that need this power. The advantages above set forth are such as Mathematics is peculiarly fitted to give, and without which they are scarcely ever attained at all. In so far as the physical sciences unfold similar advantages the way is paved for them by Mathematics.

To this short sketch of what Mathematics does, we should, for the sake of clearness, append what it does not do, and must be left undone, if we stop with it. It does not teach us how to observe, how to generalize, how to classify. It does not teach us the prime art of Defining by the examination of particular things. It guards us against some of the snares of language, but not all; it is no aid when statements and arguments are perplexed by verbiage, contortions, inversions, or ellipses. It is not the same as Syllogism in Logic, and is not in any sense a substitute for Logic, although it is a valuable adjunct. The too exclusive devotion to it gives a wrong bias of mind respecting truth generally; and, historically, it has introduced serious errors into philosophy and general thinking.

The Experimental and Inductive Sciences.

When we leave Mathematics, pure and applied, including a considerable part of Natural Philosophy, we enter the domain of EXPERIMENTAL and INDUCTIVE SCIENCE, throughout the whole of which a common character prevails in all that regards intellectual discipline. The experimental branches of Physics or Natural Philosophy, the whole of Chemistry, and Physiology, display the Experimental and Inductive Methods in their purity.

Throughout this wide field the precautions for arriving at truth by Observation and Experiment attain their highest exemplary force. The ascertaining of a solitary fact, which the untutored mind regards lightly, is in these sciences regarded as a serious task. To find out the change of bulk in oxygen gas, when converted into ozone by the electric spark, Dr. Andrews repeated one experiment several hundreds of times.

With the determining of facts goes the process of Inductive Generalization, of which these departments afford the best models. It is in this school, if anywhere, that the natural tendency of the mind to over-generalization is corrected. The history of physical discoveries is a perpetual warning against generalizing too fast; and the logic of these sciences provides the texts and canons of sound procedure. The establishment of the law of gravitation by Newton was a grand lesson in the generalizing process. The difference between an established induction and a temporary hypothesis was there clearly exemplified, and will not be forgotten. It is from the sphere of the Physical Sciences that Inductive method has been transferred to other subjects, as Mind, Politics, History, Medicine, and many besides.

In the same field we are taught when and how far empirical and limited generalities are to be trusted. We also get a practical exemplification of the rules of Probable Evidence, whose foundations are laid in Mathematics. In this, as in other points, the Physical Sciences are the best transition from the abstract for-mulas of Mathematics, with their demonstrative cer-tainty, to the regions of probability, as exemplified in human affairs.

These are but a few of the more prominent lessons of Method imparted through the physical sciences. It would take a long chapter to show how the ideas obtainable from them permeate into other fields of knowledge, as was exemplified in the Mathematical sciences.

In the point of view of information for direct use, the three subjects under consideration are by empha-sis the region of 'Useful Knowledge.' From Physics, from Chemistry, from Physiology, flow innumerable streams of fertilizing information, diffusing themselves in all the arts and conduct of life. Not only are they at the basis of many special crafts, but they provide guidance to every human being in endless variety of situations. For some kinds of knowledge we can trust to a skilled adviser; but every denizen of the globe needs perpetually to apply physical, chemical, or physiological laws, in circumstances where no adviser can be near. Still more is the occupier of a tenement in modern civilized life dependent on a ready knowledge of the truths in these departments.

The applications of Natural Philosophy to all our familiar operations are very apparent. Among our household tools are levers, pulleys, inclined planes, and many other forms of solid machinery. We have to manage windows, grates, bells, clocks; we have to consider adequacy of support in endless forms. We have, in the circulation of water, hydrostatic and hydraulic principles to carry out. We have gaseous operations in the admission and the egress of air, in warming and ventilating, and in the use of coal-gas for illumination. The principles of heating are encountered in steam tension and the explosion of boilers. It is not enough to be able to call in workmen, when anything is deranged; we ought ourselves to understand the operation of all the forces, so as to take the right precautions at every moment; and this we may do partly by empirical knowledge, but still better by the aid of scientific principles.

The immediate uses of Chemistry are perhaps fewer in number, but they are equally important. The corroding effect of acids and of alkalies, the solvent action of spirits of wine and of oil of turpentine for varnished surfaces that are unaffected by water, the protection of dresses and of furniture from dangerous chemicals used in household work, as well as many things connected with washing, with cookery, and with the keeping of household stores,—involve a certain amount of chemical knowledge.

The use of Physiology in the preservation of the health and soundness of the body, gives an additional value to the preparatory Physics and Chemistry without which it is but imperfectly understood. Although the more important results of Physiology are embodied in practical measures as to the need of pure air, sufficient and wholesome food, alternation of exercise and rest, the dependence of the mental powers on bodily conditions,—yet these great maxims are scarcely apprehended in their full force without some familiarity with physiological science. Moreover, although nothing less than professional medical skill suffices for the greater part of the derangements of the body, nevertheless, the co-operating intelligence of the patient is of the greatest help in the process of cure. But as the science is still imperfect, even in the best hands, we must not exaggerate its powers. What it does is sufficiently important to reward the study; yet to say, as has been said, that it is capable of prescribing the proper moderation of the sexual appetite, is to claim a result not as yet attainable by any science.

The three experimental sciences now named cover a very large field of phenomena, and the understanding of them enables us to penetrate into many of the secret workings of the natural world. The gratification of enlightened curiosity afforded by them is among our first-class pleasures; and their past history, with the daily record of their advancement, is a sensible contribution to the occupation of the mind and the zest

of life. The intercourse with our fellows that is based
on the giving and receiving of knowledge is the least
tainted with what is gross and grovelling.

The Sciences of Classification.

The third great scientific region is what is commonly
called Natural History, represented by Mineralogy,
Botany, and Zoology, whose peculiarity is to create a
SYSTEM OF CLASSIFICATION for embracing an enormous detail of objects. All these branches have their
other aspect as sciences of Observation, Experiment,
and Induction; they are, in fact, the previous sciences
over again, but accommodated to the emergency of
putting into orderly array the vast multitude of minerals, plants, and animals.

Now, to learn to classify is itself an education. In
these Natural History branches, the art has been of
necessity attended to, and is shown in the highest state
of advancement. Botany is the most complete in its
method; which is one of the recommendations of the
science in early education. Mineralogy and Zoology
have greater difficulties to contend with ; so that where
they succeed, their success is all the greater.

Much of the subject matter of the sciences of Physics,
Chemistry, and Physiology is agreeably repeated in the
descriptions of Natural History : a mineral is given as
possessing mathematical, physical, and chemical properties ; each animal possesses anatomical structure and
physiological function.

There is a great mass of useful knowledge mixed
up with these sciences, although perhaps more for the

special arts than for universal application. But the *interest* excited by the concrete detail is very great ; it is the easiest of all forms of scientific interest. People can be got to study and collect animals, plants, and minerals, without going deep into the physiological and physical laws. Indeed, the maximum of interest often attaches to the minimum of science, as in the search for plants ; but this taste is both something in itself, and also the introduction to more genuine studies.

In the discussions of the present day, as between creation and evolution, a knowledge of plant and animal structure is a preparation for judging of the arguments on each side. The enlarged views of recent years on the spread of vegetation lend a high cosmical interest to botanical knowledge.

Zoology is a handmaid to Human Anatomy and Physiology, on which must ever converge the highest of all utilities.

Whoever has made a study of the mother sciences— Physics, Chemistry, and Physiology—is capable of entering upon the corresponding Natural History sciences, although no single mind can exhaust the detail of any one of them. It becomes, therefore, a nice point of teaching to select some adequate representative particulars, so as not to waste time upon an interminable region of facts. The *Method* should be thoroughly conceived ; for in all studies of detail—Medicine, Law, Geography, History—that will be found to operate in imparting lucid arrangement. Indeed, clearness in style and composition depends . as much upon the arrangement of the ideas as upon the mode of expressing them, and no one subject is more suggestive of good arrangement, even in the order of a paragraph, than the method embodied in the Natural History sciences.

From the Natural History sciences we might proceed to the consideration of Geography, which has still more of the characteristics of concreteness and comprehensiveness. As it draws contributions from nearly every science, it seems to comprehend them all. This gives it a factitious and misleading charm, as if it were the grand portal to the sciences. More soberly measured, it contains a large store of practical information, it fills the imagination with vast, various, and interesting views, and it is the essential groundwork of the study of history.

The Science of Mind.

Of the fundamental departments of knowledge, I have not yet spoken of the MIND, which is explained in a separate science, called Mental Science, or Psychology.

It is generally allowed that some knowledge of the constitution of the mind is desirable. But this is seldom sought for in the science of the mind; people are content with the knowledge that comes to them in other forms; as in personal experience, in common maxims, in history, oratory, romance. All this may be good or bad as information, but it is nothing at all as method or training. In point of fact, much of it is incorrect and wrong; and the purpose of a science of mind is to rectify all that.

The student should go to the science of Mind prepared by the discipline and the information gained in the previous sciences, more especially the Mathematical and the Experimental groups. If pursued on this basis, Psychology will superadd a discipline of its own, while extending the quantity and improving the quality of our mental knowledge.

Some of the greatest problems that can occupy the attention of mankind are grounded in the human constitution ; and the scientific handling of mind has been often impeded by the partisan solutions given to such questions as Absolute Being, Innate Ideas, the Moral Sense. Unless entire impartiality can be shown in dealing with these subtleties, a theory of the mind may darken all that it touches.

The subject of LOGIC is usually associated with the Science of Mind, although it has an independent standing of its own. Logic, in the enlarged view of it taken at the present day, is a suitable accompaniment of a course of the Sciences, as we have sketched them. It directs attention upon the points of Method or Discipline in each science, which the cultivator is apt to neglect in his zeal for the matter or information of each. Even with Mathematics, a logical commentary is desirable ; it is no less useful both in the Inductive and in the Classificatory Sciences.

The foregoing sketch comprises the field of the theoretical or knowledge-giving sciences, those that embrace the most complete and systematic view of all the kingdoms of natural phenomena. They both present the scientific method and spirit in the greatest perfection, and impart the greatest amount of accurate information. Whatever scientific culture can do, is done by the curriculum thus laid down. Of this culture, perhaps the greatest result is embraced under the devotion to TRUTH, which, allowing for human infirmities, must emerge as a consequence of being initiated in all the devices of modern research. How the cultivation of this cardinal virtue tells in every department of life need not be here insisted on. The moral disposition to veracity avails little without the tests and methods of distinguishing true from false, while men well versed in these seldom quarrel on matters of fact, seldom keep up irritating controversies as to what is or is not. The disputes of the scientifically educated are narrowed to some very special and difficult issues.

The Analyzing operation, which pervades all science, is most pointedly opposed to the crude and clumsy procedure of the untutored mind, which insists on treating things in the lump. The British Constitution is a vast mass of arrangements, which a scientific politician views separately, pointing out which are instrumental to our safety and happiness, which are detrimental and need amending, and which are neutral or indifferent. The vulgar reasoner will speak of the mass only as one indivisible agent.

The bearings of Science upon Fine Art should be properly understood. In the first place, Science checks the extravagant departures from truth in Art, and is thus a medium of purifying Art productions. This is a great negative result ; for it is an undoubted tendency of Art to depart from truth in order to pander more fully to ideality and illimitable desire.

In the next place, Science discloses new facts, new

laws, new views, which have more or less power to interest our feelings, and thus become materials for the artist. Astronomical discovery has furnished many new and enlarged conceptions of the celestial sphere, tending to nourish the sentiment of the highest sublime. The terrestrial forces have been displayed in novel and striking aspects through the physical discoveries; and the result is both to cultivate poetry, and to make science itself poetical.

In the third place, it is not to be concealed that Science and Art pursue totally different lines, even to antagonism. The analyzing operation of science is at variance with the concreteness of poetry; the abstract, uncouth, and technical expressions that grasp scientific truth, are repugnant to the artistic tastes; the check interposed to poetic ideality by the severity of scientific truth, is an abatement of our artistic pleasures.

Striking a balance among these three considerations, we conclude that the artist should have a certain amount of scientific education, as a prelude to his art, while he is not to be expected to keep his mind immersed in the scientific ideas and forms most removed from æsthetic culture. Two of the most luxuriant imaginative minds of this century—Thomas Chalmers and Thomas Carlyle—were in youth good mathematicians; and much more might a man of artistic mould afford to drink deep in the Inductive, Classificatory, and Mental Sciences.

The Practical or Applied Sciences.

In these, the matter of the Knowledge-giving sciences, as above enumerated, is turned to account in

practice, and is disposed with that view. In the practical science of Mensuration, the propositions of Euclid,
the rules of Arithmetic, and the formulæ of Algebra, are
torn from their context in a Mathematical system, and
exhibited in the order suited to the questions in hand.
In such sciences, the connected form of science is absent
the information imparted is much less compared with
the bulk ; the needs of the practical man are alone
considered.   The practical sciences of Navigation, Machinery, Engineering, Metallurgy, Agriculture, Medicine
and Surgery, War, which are all in relation to Physical
Science, must remain as the special acquisitions of professions or crafts.   The practical departments related
to the Human Mind, as Politics, Ethics, Law, Grammar,
and Rhetoric, are of more widespread interest ; a large
portion of them entering into general education.   On
these a few remarks may be made.

And first, of the Sociological group—including Politics, Political Economy, Legislation, and Law or Jurisprudence.   Politics is the science of Government, so far as
regards the form of government—whether Monarchcial,
Aristocratic, or Republican.   It is in close alliance
with History, whose highest aim is to throw light upon
the constitution and workings of Government, in which
lofty aim it aspires to be an independent branch of
study, called Historical Philosophy, or the Philosophy of
History.   This subject is still in a somewhat unsettled
state, although rapidly tending to become organized
under the name of Sociology.

Political Economy is a separate department of
Political Science, concerning itself with the laws that
regulate Industry to the greatest advantage.   Its place

in Education is, therefore, considered very high among practical sciences. To a student accustomed to scientific reasoning, it is not a difficult subject : still it is one that needs the aids of public teaching. To promote an enlightened opinion in matters regarding Trade, all educated persons ought to know something of this department; while, in the operations of Government, it is imperatively required. It lends indirect support to the moral habits of industry, justice, and veracity, in which character it ought to be universally diffused; but in that case the teaching should be so conducted as to make these lessons prominent.[1]

Legislation in its widest sense includes all the operations of the supreme Legislature; but a portion of these relate to the constitution of the Government, or Politics in the narrowest sense ; while another portion comprises the laws relative to Industry, as grounded in Political Economy. The prevention of Crimes and the enforcement of Rights constitute a large department, including Penal Legislation or Punishments. Legislation also determines all the relationships of Family ; the conditions of Service ; Pauperism ; Education ; the relations of the State to Religion. No one science includes the whole of these topics.

Law or Jurisprudence, which are nearly the same thing, is a limited subject connected with the *form* and expression of the Laws, as distinct from their substance. It teaches how laws should be codified, so as to be compact and intelligible ; and how they should be worded,

---

[1] Mr. William Ellis has long distinguished himself in this application of Political knowledge. See his *Outlines of Social Economy.*

so as to admit of exact interpretation. It embraces Evidence and Procedure.[1]

Ethics is a science of such conflicting views that, as far as concerns its foundations, it is included in the higher education, being usually associated with Mental Science. Its preceptive part belongs to the diffused knowledge of the people, and is inculcated at every stage of life, making up what is called Moral Education.

The sciences of Language are Grammar, Rhetoric, and Philology ; the two first have for their subject the immediate employment of speech with propriety and effect ; the third, General Philology, takes a higher speculative sweep, and is one of the subjects involved in the historical evolution of the race. Each Language has its own Grammar, which is taught with the language. Rhetoric lays down principles applicable to all languages,

---

[1] In a recent address by Sir James Fitzjames Stephen a claim is made for Law, as a branch of public education : 'It has often been to me a subject of great surprise, that while the slightest alteration in the machinery by which laws are made excites intense interest, the laws themselves, when they are made, are treated not as a subject of liberal study and education, but as a mystery known to only a few students, and incapable of being communicated to the world at large. I have long been of opinion that such subjects as the criminal law, the law of contracts, and the law of wrongs are in themselves quite as interesting as the subject of political economy ; and I think that if the law were thrown into an intelligible shape, the result would not only be of the greatest possible public convenience, but would constitute a new branch of literature and of public education.'

There can be little doubt that Law is a most valuable discipline in many important matters connected with everyday life.   It tends to arrest precipitate conclusions as to the guilt of supposed wrongdoers, and promotes justness of dealing in our relations with others.   If these lessons could be extricated from the load of details necessary to the professional lawyer, and presented in a brief compass, they would deservedly rank as a liberalizing study.

Bentham considers that the People should be so far instructed (in political matters) as not to quarrel with their own interest.

although with some special modifications for each; an inflected and uninflected language cannot employ the same devices of arrangement of words in the sentence.

These various Practical Sciences have no purpose beyond their immediate application. None of them can be accounted sciences of Method, Discipline, or Training. The opposite view is held by many with reference to Grammar; and the arguments will be afterwards considered. In the meantime, we lay it down that these sciences, by being exclusively accommodated to their practical objects, do not, as matter of course, set forth the arts and devices of science to advantage; they repeat in an inferior shape what is best given in the fundamental or knowledge-giving sciences. As branches of practical knowledge, they ought to be exact in their statements and supported by adequate proofs; but they do not lay themselves out for giving instruction in the principles of evidence.

LANGUAGES.

We now turn to the great field of *Language*. While the Mother Tongue is an indispensable acquisition, an interest also attaches to the Languages of other nations, so much so as to lead to the including of these in our regular course of Education.

The learning of a Language has a value according to the use that we are to make of it. This is admitted. If we are to listen to French, speak French, read and write French, we must be taught the language. So, Latin being the literary medium of the Middle Ages, had to be known by every scholar. But if we are not to use

13

a Language at all, or very little, as is the case with the majority of those that learn Latin and Greek at school and college, is there any other reason for undergoing the labour?   This is the question of the day, as to the utility of the Dead Languages.   At a later stage I will consider the arguments, on both sides.   At present I intimate my view, as regards the learning of Languages, that their main, if not their sole, justification is that we mean to use them as languages, to receive and to impart knowledge by their means.   This does not exclude the pleasure that we may take in the poetical compositions of a foreign tongue.

Language is, in the first place, a series of vocables addressed to the ear, and to the eye, and reproduced by the voice and the hand ; and these have to be associated with the objects that constitute their meaning : a very extensive exercise of memory.   Equally a matter of memory is the customary arrangements of words and sentences ; although at this point the practical science of Grammar comes into play, followed up by another science— Rhetoric.   These sciences, however, are of value only as aids to the knowledge of the language ; and, if employed upon a superfluous language, are themselves superfluous.   It is true that Rhetoric is not confined to any one language ; nearly the same precepts are applicable to all.   Yet that is no reason for connecting it with an unused language ; we can always find exercises in the languages that we are to speak or to write.

Science and Language embrace between them the great field of Intellectual Education, including also the higher parts of the education in professions and crafts

They do not comprehend, unless incidentally, Mechanical Training, the Training of the Senses, Art Training, or Moral Training. The two last, Art and Morality, will receive separate chapters ; some remarks may be offered on the two first at the present stage.

MECHANICAL TRAINING.

Mechanical Training includes the command of the bodily organs for all the ordinary purposes of life, and the special training for special aptitudes. The child's spontaneous education furnishes the commencement ; imitation and instruction follow. Mechanical training is implied in writing and in drawing, which are a part of school training ; also in the handling of tools, and the performance of operations in the various crafts ; in household work ; and in active amusements. The hand receives a special training in playing on a musical instrument. Parallel to the manual training is the vocal training in speaking and in singing ; while there is a training in gesture for elegant deportment.

It is a part of the idea of the early training of children in the Kindergarten to push them forward in the manual accomplishments, that is, to give them the early use of their hands. Irrespective of special arts, there is much difference between one person and another in manual facility, as applied to the countless little emergencies of life ; and it is a great advantage to have good hands. Nevertheless, this is not a matter for the public teacher to spend time upon, further than is required for other purposes. If children can be interested in any mechanical employment, they will acquire skill in it ; but there is an error in allowing them to be engrossed in the lower energies of the mind to the neglect of the higher.

TRAINING OF THE SENSES.

The Exercise and Training of the SENSES is much insisted on, but is not well defined.   Here, too, there is a general training suited to all, and a special training for special arts.   To train any one of the Senses is to increase its natural power of discrimination, as in colours, tones, touches, odours, tastes.   An artist in colours undergoes a training for colour discrimination ; a musician and an elocutionist possess an acquired delicacy of hearing ; a cook has a training in the palate.   This is the most precise meaning of the Improvement of the Senses.   Out of this superior discrimination will grow a better memory for the respective sights and sounds and tastes; so that the conceptive concrete faculty will be strengthened at the same time.

The early training of the Senses, as usually prescribed and practised with Infants, points several different ways.   There may be an augmented discrimination of colour; also of visible forms and visible magnitudes, so as to give a finer sense of the magnitudes and properties of objects.   This is held to be a preparation for at least three different accomplishments :—first, correct estimates of the colours, forms, and sizes of things at sight ; second, the arranging of colours and forms into symmetrical groups, to gratify the art sensibility; thirdly, the understanding of the figures of geometry.   The first of these accomplishments

—namely, the accurate estimate of colour, form, and size by the eye—is of little use generally ; it applies to the special arts, and particularly drawing and design, with which it is necessarily bound up. This, too, is the meaning of the second accomplishment, which is carried to marvellous lengths in the Kindergarten ; the children being led on to contrive and to execute elegant symmetrical forms, by grouping simpler figures in innumerable ways. This should not be called Sense Training ; it is a special education in Drawing and Design. As to the third end—the preparation for Geometry—there is nothing to show that this needs any such training or depends upon it. The sense foundations of Geometry are so few and simple that no one can well escape them ; and the science speedily and peremptorily demands that the senses shall give place to the constructive reason. A geometer must not mistake a triangle for a square, or a circle for an ellipse, but he does not need a delicate visible perception of the exact proportions of the ellipse; he never depends upon the eye for a measurement ; he need not be able to detect by sight a small deviation from the perpendicular.

The utility of DRAWING as a general accomplishment must not be overrated. It is an additional acquirement of the hand, and for special purposes is valuable or even indispensable. But, as a foundation of intellectual training, its influence is liable to be mistaken. It is supposed to train the observing powers, thus helping to store the mind with the knowledge of visible objects. But this is too vague to be correct. Drawing compels the child to observe just what is necessary to the end and no more : if to copy another drawing, the lines of that must be carefully noted ; if to draw from nature, the form and perspective of the original must be attended to : but this does not imply much ; it does not involve an eye for outward things generally in all their important characters. The pupil does not necessarily give any more heed to the things that he does not intend to draw. Observation, in its full meaning, is not a matter of the senses purely; it consists in interpreting indications, by applying previous knowledge, and is a special training within a limited sphere. Such is the observation of the Astronomer, the Geologist, or the Physician.

When Drawing is pursued so as to become a taste and a fascination, it is too engrossing ; it disturbs the balance of the mind, and indisposes for other tasks. Worst of all, instead of leading the way to science, by assisting to stamp on the understanding the pre-requisite assemblages of particulars, it resists the farther advance from particulars to generals, and it clothes the particulars with such a degree of concrete interest, that the mind prefers to remain in the concrete. A moderate taste and aptitude for drawing may be helpful in the more concrete sciences ; especially, if it goes no farther than drawing. But when the colour interest takes a deep hold of the mind, it imparts a too exclusively pictorial character to the intellect, and breeds an unfitness for the abstract and analytic procedure of science.

CHAPTER VI.

SEQUENCE OF SUBJECTS—PSYCHOLOGICAL.

THE long chapter on the Psychological basis of Education leaves one point untouched—the Sequence or Succession of the Powers and Faculties of the Mind. It is important for us to grasp, if we can, not merely the leading components of our intellectual structure, but also the order of their unfolding.

If we could suppose the brain, at birth, to possess all the physical capabilities of our brain at twenty-one, but a *tabula rasa* in respect of impressions of every sort, the order of acquisition would be the strict order of dependence of one thing upon another. Simple elementary impressions would come first, and would be followed up by those of a more complex kind ; the concrete would precede the abstract, and so on. Priority of study would follow a very plain rule ; Analytical or Logical sequence would be the one principle of Order. The actual case, however, is very different.[1]

---

[1] Anatomists tell us that the brain grows with great rapidity up to seven years of age ; it then attains an average weight of forty ounces (in the male). The increase is much slower between seven and fourteen, when it attains forty-five ounces ; still slower from fourteen to twenty, when it is very near its greatest size. Consequently, of the more difficult intellectual exercises, some that would be impossible at five or six are easy at eight, through the fact of brain-growth alone. This is consistent with all

The fact that the educator works upon a growing brain, and not upon a completed one, does not invalidate the law of logical order ; it only imposes another set of

our experience, and is of value as confirming that experience. It often happens that you try a pupil with a peculiar subject at a certain age, and you entirely fail ; wait a year or two, and you succeed, and that without seemingly having done anything expressly to lead up to the point ; although there will inevitably be, in the meantime, some sort of experience that helps to pave the way. In regard to the symbolical and abstract subjects, as arithmetic, algebra, geometry, and grammar, I think the observation holds. A difference of two or three years will do everything for those subjects.

This, however, is but one aspect, although a very important one, of the varying rate of brain-growth. If we follow the analogy of the muscular system, we shall conclude that the times of rapid growth are times of more special susceptibility to the bents imparted during those times. If the brain is still unable to grapple with the higher elements, it is making great progress with the lower ; whatever it can take a hold of, it can fix and engrain with an intensity proportionate to its rate of growth. That is a good reason for looking well to the sort of impressions made upon the child during the first seven years.

It would be a great contribution to our subject, if we could fix with any degree of definiteness the variation of the plastic adhesiveness of the brain through life ; beginning in those years of infancy when it is greatest, and going on to its extreme deficiency in old age ; the decrease being, I should presume, steady after some year between six and ten. But the determination is full of difficulty, owing to the number of collateral circumstances that obscure the main fact.

The growth of the brain is no doubt accompanied by the perfecting of a number of innate powers, without which our education would be something totally different from what it is. That many of our notions of the outer world have the way prepared for them by hereditary impressions, or instincts, is a received doctrine of the present day. How far this is so, we cannot precisely estimate. For practical purposes, we must observe the total appearances presented us in the growth of the infant mind ; we cannot disentangle what depends on brain-growth, with hereditary transmission, from what is due to contact with the actual world. We must be content with noticing, as a matter of fact, the age when abstract notions can be taken in, without deciding whether the growth of the brain, or the accumulation of concrete impressions, is the principal antecedent.

conditions. I will make a few imaginary suppositions by way of illustrating the real state of the case.

As one alternative, the immaturity of early years might amount to positive defect in an important sense organ, say sight or hearing. In that case there would be a blank in certain impressions that are indispensable as an ingredient in some department of knowledge. Imperfect discrimination of colours would arrest the knowledge of the object world ; while a want of the sense of form would be still more fatal. A very large part of our education would thus be in total abeyance.

Again, the sense capabilities might exist, but in such an imperfect degree for the first few years, that it would be bad economy to attempt to found upon them. The natural course of growth might be such, that by waiting a year or two, acquisitions that are attended with difficulty at an early stage, could then be accomplished with ease.

Thirdly, there might be intellectual susceptibility as regards all the essential properties of natural things, but a want of the vigour of intellectual attention. To strain the powers at this stage would then be sheer waste, in consequence of interfering with the physical growth.

Or, lastly, there might be susceptibility and even a certain amount of power of attention, but an immaturity of the requisite motives and interests. The feelings and dispositions might be for a time alien to everything intellectual, being engrossed in sense pleasures and excitements, whose results as regards knowledge could only be accidental, random, and desultory—Imagination would be preferred to fact.

These four suppositions all correspond in some de-

gree to the reality. We do actually postpone the commencement of many studies because the mind at an
early stage cannot entertain even their most elementary
conceptions, and still more because we cannot procure
the requisite attention, from want of routine and persistence. We do not in this, however, exclude all reference to the analytic or logical priority ; inasmuch as
there may be wanted a certain spontaneous absorption
and fixity of sense impressions before any training operations could be commenced.

The first approximation to defining the order of the
faculties is given in the commonplace remark that observation precedes reflection ; or, in another form, that
the concrete comes before the abstract; which is good so
far, but not very precise. Another maxim is, that the
Imagination is an earlier faculty than the Reason ; this
too needs qualifying.

As an example of the questions to be settled by the
present enquiry, I may refer to what is a suitable age
for commencing study, as typified by learning to read.
The practice and the opinion on this head show a wide
disparity ; the range is from three years old to seven.
Further questions arise with respect to the proper times
of commencing Languages and Sciences respectively.
On this head, there have been differences of view.
Language, being chiefly dependent on Memory, would
seem to come early, as Memory is strong, while Reason
or Judgment is still weak. The commencement of
Science needs not merely a preparation of concrete facts,
but an advanced form of interest or emotion, and a
great control over the mental attention, which is a late
acquirement.

There is an additional point of nicety, and yet of importance, namely, to assign the commencement of self-conscious or subjective knowledge—the facts of the incorporeal world ; many of which have to be taken for granted in the earliest species of composition addressed to the young.

Let us first survey the peculiarities of the infant mind ; and, in so doing, we shall sketch the earliest steps in knowledge, as depending on those peculiarities :—

Every one of us has watched, with more or less attention, the mental phases of human beings at their different stages of growth ; and the result has been a certain vague estimate of the changes that advancing years produce upon the faculties. But, in proceeding to render an account of the actual sequence, we are met at once with the difficulty of finding terms suitable to describe our observations. There are a few set phrases that are regularly brought into the service. The child, it is said, has a great love of activity, a desire to be occupied somehow ; dislikes continuous application or attention to any one thing ; is joyous, mirthsome, fond of fun and frolic ; delights in the exercise of the senses, and in sensation generally ; is curious and inquisitive, even to destructiveness ; is strongly given to imitation ; is remarkably credulous ; is imaginative and fond of dramatizing ; is sociable and sympathetic. On the more exclusively intellectual side, the child is prone to observation, and averse to abstraction ; is strong in memory, and weak in judgment

To reduce these observations into order, we must bring them under the usual classification of mental elements—Activities, Senses, Emotions, and Intellectual Powers. First, then, as to ACTIVITY. This is spontaneous and abundant, but fluctuating, uncertain, and indirect, being the outpouring and overflow of natural energy. Among the first efforts at education is the attempt to give it useful directions; but the readiest way is not to force it, but to take it at the moment when it has fallen into a good course. Second, as to the SENSES. These being fresh, and everything being new, sensation as such is delightful and coveted; hence, the employment of the senses, and the fruition of the effects, are intense in infancy. But, at first, the emotional side preponderates, and the intellectual side, which is nourished by nice distinctions, does not attain an early development. The emotional force partly paves the way for, but partly obstructs, the intellectual. Third, the EMOTIONS, strictly so called, as distinguished from the sense enjoyments. These are mainly the strong social feelings—Love and Affection; the strong anti-social feelings—Anger, Egotism, Domination; together with the workings of Fear. All are powerful from the dawn of life; education, while connecting them with special objects, may do something to intensify or to enfeeble their total force. Fourth, the INTELLECT. The fundamental tendencies or functions—Discrimination, Discovery of Agreements, Retentiveness or Memory, are at work from the first; but the active emotional development keeps them all down at the outset, although doing something to provide materials that will be used on a future day. The operation of intellect is requisite to such complex growths as curiosity, imagination, dramatizing, imitation, and fancy. The higher workings of intellect become necessary even to the observation of facts in any form that deserves the name.

The management of the Activities, the Sense Pleasures, and the Emotions, makes up the branch of education called Moral Education. As essential forces and adjuncts they are all taken into account in intellectual education, for which department, however, the principal thread must follow the growth or sequence of the intellectual powers.

The beginnings of knowledge are in activity and in pleasure, but the culminating point is in the power of attending to things in themselves *indifferent*. The successive stages of the process may be conceived as follows:—

By common consent the first start in knowledge is made through spontaneous and overflowing activity and the interest of the impressions of the senses; all which, in the pristine freshness, afford an abounding enjoyment. At this stage, many things are discriminated, and from discrimination all knowledge begins. But then to discriminate is not a primary vocation of the infant mind; enjoyment—immediate and incessant—has the precedence of all other objects. In the presence of the more enjoyable, the less enjoyable is disregarded. Observation, attention, concentration, lasts so long as enjoyment lasts and no longer. When the interest in anything flags, something else is sought after. If the pain of attention is greater than the pleasurable excitement, the attention is withdrawn. This state of things is so far favourable to knowledge; a good many objects ot sense are surveyed under the pressure of pleasing attraction; restless activity leads to many changes of view, and the search for excitement induces a repeated survey of the outward scene. Moreover, *intensity* of sensation, whether pleasing or not, is a power; this does not win by seductive charm; it takes the attention by storm. Things indifferent, and even things unpleasing, leave their impress by the severity of the shock they give. There is an old saying, that wonder is the beginning of philosophy. Various things may be meant by wonder, but one thing is the shock of mere surprise or astonishment, irrespective of pleasure imparted. If the shock is painful, the mind no doubt rebels; it perhaps goes off in search of some sweet oblivious antidote; but an impression has been made—an element of knowledge is secured.

Before discussing the transition from the experiences that impress on the mind what is pleasurable, painful, and intense, to the impressing of those things that in themselves are indif-ferent and insipid, which make the larger part of our knowledge in the long run, I must bring up the side of activity to the same point as the side of passive receptivity. The active energies in the first instance follow the same course of adhering to what

gives attraction or charm; at all events they do not lend themselves to tasteless effects. The moving organs, as repeatedly
observed, begin by being exercised under the pressure of the
active centres, their exertion being determined and limited by
the central energy. When the steam is expended, the action
ceases. A certain pleasure goes along with the active expenditure, but the action and the pleasure cease together, when
the nervous and muscular discharges are no longer maintained.
Under this prompting, the movements do nothing that is useful,
except by accident; they do not of their own accord fall into
any of those combinations that serve some productive end.
No doubt they are preparing for such combinations. We cannot suppose that the child moves all its limbs profusely and
variously without both strengthening them individually, and
enlarging their compass or sweep; in short, bringing them up
to the point when they can enter into groupings for useful ends.
I am not here inquiring into the precise limits of the instinctive
and the acquired actions of childhood. It is enough to recognize the fact that the first useful combinations are accidental;
the discovery of their use is the cause of their being maintained,
continued, and ultimately fixed into habits and active capabilities. In a word, pleasure and abatement of pain are the first
motives to acquirements in the bodily organs. The power of
the hands to supply wants, cater for pleasures, and rebut pains,
is the earliest manual aptitude. The motions of head, trunk,
eyes, mouth, tongue, all come into the like service, and this is
their earliest stage of culture.

Of all our muscular aptitudes, the most illustrative is Articulate Language. At first purely spontaneous and emotional, it
lends itself very speedily to our desires and purposes, and in that
service receives the commencement of its cultivation. The tones
that demand assistance, that express satisfaction, or the opposite, become detached from mere instinctive promptings, and
pass into useful instruments of the various moods and wishes
of the infant. Then comes the child's pleasure from hearing
the sound of its own voice; in which case it will cling by pre

ference to the more agreeable tones (according to its standard
at the time).   But most illustrative for our purpose is the early
stage of imitation—the stage when it is a pleasure to reproduce
the sounds made by others.   The motive here is somewhat
advanced and complex, and does not put forth all its power
till a later period; but it exemplifies that primary stage when
nothing is done without some immediate gratification.   The
social instincts are undoubtedly very early in their appearance;
and one of their manifestations is the interest felt in personality
as such, and beyond the mere utility of being fed and attended
to.   The infant soon shows a degree of engrossment with persons that transcends the supply of its primary wants, although
involving these; and this interest makes the charm of imitation.
Having given voice to an articulate sound heard from others,
the child experiences a throb of delight from the coincidence;
and such pleasure is the early support and stimulus of imitation.
It adheres to us all through, and is one of the teacher's best
aids.   Disgusted, as he often is, to have to cram things down
the throats of unwilling subjects, his work is now and then
lightened by the operation of this motive to imitate and reproduce with alacrity his own special aptitude and skill.

To come now to the second stage of culture—the acquisition
of the INDIFFERENT, both as passive impressions and as active
power.   We cannot be too thorough in our study of this critical
transition; it is equalled in importance, but not surpassed, by
one other transition, namely, from the Concrete to the Abstract.

To escape from the influence of pleasure and pain as motives is impossible.   To fall in love with and pursue the indifferent and insipid is a contradiction in terms.   It is as *means to
ends* that things indifferent in themselves can command attention.   We may have the capability of distinguishing minute
differences in the lengths of two rods, in the weights of two balls,
in the curvatures of two bent bows, in the shades of two reds,
in the pitches of two notes—but if the act gives no pleasure,
removes no pain, excites no astonishment or violent sensation,

we decline the exercise.   By the first law, the prime condition, of all consciousness, a *considerable* difference has an awakening power, something of the nature of a surprise, and it leaves an impression which becomes an item of knowledge.   A sharp change in the light of a room, a sudden rise or fall in the intensity of a sound, awakens the consciousness; and the more delicate the sense, the smaller are the changes that impart an arousing shock.   This is the length that we can go in impressing mental differences.   But I apprehend that the agency of difference, as an awakening or a shock, is very far short of our capability, as well as our needs, in the way of discrimination. Passing from one room to another ten degrees hotter or colder, we are aroused to the difference whether we will or not : perhaps five degrees might give the awakening ; but it needs the pressure of some special motives to attention to make us discriminate (as we are able to do) a transition of one degree.

One of the first indications of growing intelligence, of the contracting of fixed impressions of things around, is the discovery of circumstances *attendant on what gives pleasure* ; events and objects that precede or accompany things that are delightful in themselves.   The stimulus to attention derived from what is agreeable operates towards these accompaniments, which are thereby discriminated, marked, and impressed on the memory.   The child comes to know, not merely its food and its agreeables, but all that goes along with them, and all the prognostics of their arrival.   An object of strong intrinsic interest irradiates its surrounding sphere, and the more so as the impressions of outward things harden and become coherent.   In this way great additions are made to the stock of discriminated and remembered objects ; the motive being still an interested one—the access of pleasure and the avoidance of pain.   The motives continue the same, but they are intellectually extended.   The wider the view of the collaterals of our pleasures, the wider is the influence of the stimulus to attention and discrimination.   A very faint sound, which as pleasure is nothing, as a shock is unheeded, may yet betoken the arrival

of some welcome person or some known gratification, and as
such it is felt and noted.   Slight may be the distinction to appearance between the cup of genial slop, and the cup that is
doctored with extraneous matters, but that slight distinction
receives an indelible stamp.

But now we must add another consideration that leads us a
little further into the sphere of disinterested attention.   In the
absence of any strong interest. the active senses cannot help
disporting themselves for a time upon what they can get.   They
will not remain longer over a dull job than they can possibly
help; but when they cannot do better, they will take up with
what they find.   These intervals between the stronger excitements are favourable to the noticing of unattractive objects and
smaller distinctions.   The child at first is struck perhaps only
with a glaring colour—a strong red or blue, or a mass of several
shades, which it receives as a total effect.   Should familiarity
blunt the interest to the strong effect, in the absence of fresh
attractions, the mind may re-occupy itself with the aggregate of
colour, and be awakened to the distinctions of the shades.   To
discover a difference is not, in early years, an exciting employment; there is much more stimulus in the discovery of agreements: yet the exercise of the mind in bringing out any new
effect whatever brings a reward to the childish sense of power.
The moral of this line of remark is against pampering and
over-exciting the infant mind.   What is the vaunted joyousness of children if it does not mean that they maintain a cheerful glow on few stimulants; that a mild interest can satisfy
them, and leave the attention free to scan the less exciting
features of the scene, so as to gather in the minuter distinctions that widen the basis of knowledge?

So far we have regarded the child as self-prompted and self-
acting, and have endeavoured to trace the expansion of the
intellect under the motives that we suppose to be at work.   In
passing now to the artificial direction of the attention through
the influence and dictation of others—the schooling, properly
so called—we have the same motives at bottom, with a change

14

in the manner of applying them : the facilities and the precautions are still the same. A demand is now made that attention shall be paid to a class of distinctions hitherto overlooked ; to the difference between two, three, and four, to varieties of shades of colour, of articulate sounds ; and, at last, to those most minute and uninteresting differences between the visible forms called letters. Not immediate or mediate pleasure, not startling surprise, not intrinsic attraction enough for the dullest of vacant moments, would procure attention to such things, and far less a strenuous attitude of concentration ; and there is no force available but the *sic volo* of the superior. What, then, on general principles, is the most expedient course, in order to be as mild, and yet as effective, as possible? In the first place, the instructor has to establish his or her influence on the best possible foundations, by a hold on the child that dispenses as far as may be with terrorism. This we all grant. Then the natural workings, as manifested in the prior stage, should be so far attended to, that self-sustaining interests should be awakened when they can. This too is granted. Then comes the stern conclusion that the uninteresting must be faced at last ; that by no palliation or device are we able to make agreeable everything that has to be mastered. The age of drudgery must commence ; every motive that can avert it is in the end exhausted. What then? Try to measure the child's power to support the strain of forced attention. Use this power to the full, without abusing it, if you can judge the happy mean. Begin the discipline of life by inuring gradually to uninviting, to repugnant and severe occupation ; but see also that you have at command the alternative of relaxation with enjoyment.

Let us now advert to the questions growing out of the order or development of the Faculties.

At what age should education commence? We commence too early, if we interfere with the powers needed for growth ; and even supposing this does not happen, we begin too early, if the desired impressions demand much greater expenditure than would be neces-sary at a later time. On the other hand, we commence too late, if we allow time to pass by, when good and useful impressions could be made with perfect safety to the general health. This is just as possible a case as the other.

Nothing but observation of cases will avail us here. We have to set aside the instances that are extreme either in vigour or in weakness. We know that many have begun to read at three years old, and have grown up perfectly healthy and strong. What we do not so well know is whether, by beginning at four or five, they would not have been as far advanced at fifteen as they are in the earlier commencement. If, however, any considerable number of children have begun schooling between three and four, without more than an occasional instance of observed mischief, then a year later ought to be a margin of safety for all but exceptional cases. The necessity and expediency of protracting the age of commencing till six or seven cannot be made out. There ought to be proof positive that in such belated instances the child advances with a rapidity that carries all be-fore it.

At what time should we begin the mechanical train-ing of the hands, the training of the voice, the training of the eyes in observing forms and colours? We here proceed upon a natural spontaneity, which needs directing and coercing; the coercion being more or less painful in itself, and palatable only by the interest evoked.

A further question relates to the priority of different classes of acquisitions, as to their time of commencement: as Language, Knowledge of Things, Mechanical Apti-

tudes, Moral Impressions.   When is the child prepared
to take up these several departments without a strain
beyond its years ?   In all the kinds, there is the spontaneous, or self-moved commencement, followed by the
gradual attempt to direct it into definite channels.   The
rule seems to be that activity is always prior ; and according as an acquisition has an active element, it comes in
earlier, regard being paid to the state of advancement of
the special organs.   Language (spoken) seems the most
precocious of all acquirements ; being usually in advance
of the manual capabilities.

The activity of the eye is also very early, and the
cognition of visible movements, magnitudes, forms, and
of all the space relations, proceeds rapidly.   This is the
stage of spontaneous observation, and of impressions in
the concrete ; and is the necessary grounding for the
artificial education in things.   The pre-school education
consists in developing the articulate capacity, in cultivating an interested observation of surrounding persons
and things, and in connecting names with these various
objects.   The further these three branches have gone,
the better is the child fitted for the more methodical
instruction of the school.

The next point to be considered relates to the age
when Memory is at its best, and when acquisitions
in pure memory take precedence of others.   This applies
to the problem of Languages as against Science, that is,
knowledge more or less generalized, reasoned, and connected.   Now, it seems evident that for the years between six and ten very little can be done that involves
severe processes of the reason ; and yet the mind is
highly plastic and susceptible ; so that presumably this

is the age of the maximum of pure memory, as typified by Language acquisitions, not merely vocables and their connections with things, but connected compositions, as stories, hymns, and the expressed forms of knowledge.

The easier kinds of matter of fact, where generalization is carried merely to the length of increasing the interest and lightening the memory, such as geographical particulars and plain narratives, appeal more to the memory than to any higher power, and belong to the years that I am supposing.

That the harder sciences, as Grammar, Arithmetic, and Mechanics, should be later in being understood is owing, not solely to the necessity of pre-storing the mind with instances in the concrete, but also to a defect in the power of compelling the attention to perform the necessary junctions and disjunctions of ideas ; which power must be dependent upon age, in the first instance, although it is susceptible of being forced on by the efforts of the teacher. Usually, however, the premature bringing forward of these subjects ends in their being taken up in the memory alone, which can be induced at the early age to embrace even unmeaning statements. At the height of the mental plasticity, which may be from seven to eleven, interest, although aiding, is not essential ; the consciousness of the power is enough to make it not a drudgery.

It is customary among the higher ranks of society to make use of this early plasticity in laying the foundations of foreign languages, as French and German. This is so far good ; but one can easily conceive the practice of memory-stuffing carried too far. While using the moment of greatest adhesiveness, we should also be doing something to bring forward the reasoning power, in a slow and gradual way. The 'Age of Reason' should on no account be thrown back, any more than it should be precipitated. The faculties should not be absorbed by huge tasks of committal to memory of mere words; even the conceptive power of embracing the concrete meanings may be stifled in this way, and the chances of the reasoning faculty be thus doubly impaired.

It is specially interesting to view this plastic moment with reference to *moral* impressions. Commands, maxims, verbal directions are all well laid up in the memory; even the more difficult doctrines of religion may find a lifelong lodgment by being iterated between six and ten. All this, however, is external to conduct. We must look at the dispositions to obedience, the culture of the affections and sympathies, and the foresight of remote consequences. Now, as regards obedience, the discipline of fear may do much, because of the weakness and susceptibility of the subject. The other elements · are more difficult to command ; and the chief question is, whether this plastic period is favourable to pleasurable associations, assuming that the child is well supplied with things agreeable. I should be disposed to answer in the affirmative ; remarking only that this is a very costly acquisition, and may not, in ordinary cases, make much apparent way in the course of two or three years. These associations, however, rest on the same basis as the moral affections and sympathies.

As to the foresight of consequences, that is a very tardy affair. It needs a high development of the conceiving power, together with a class of associations rendered very difficult by the strength they must have attained before they answer their purpose. The opposition to encounter in the furious impulses of those early years is the measure of the strength of this association.

CHAPTER VII.

SEQUENCE OF SUBJECTS—LOGICAL.

THE previous chapter is intended to clear the problem relating to sequence by allowing for the development of the powers and faculties irrespective of the impressions that are received. We must now view the order of the impressions themselves, according to their logical dependence.

Thus, take the case of Language. The sequence consists (1) in the articulation of syllables, (2) in forming syllables into words, and (3) in uniting words into consecutive speech. So, with our conceptions of the concrete world, there is the same plain course: we proceed from elementary forms, colours, objects, views, to binary, ternary, and higher combinations of these. There is no break, or abrupt transition, at any one point. In the mechanical arts it is the same. Our state of development settles the time for beginning; when once begun the course follows the law of analytic progress. There may be an error committed in trying to go too fast; which simply means that we are taking a new step without having matured the one previous; the remedy is obvious, it is not to wait on faculty but to ply exercises.

In machinery we proceed from the constituent parts to the whole. So, the Anatomist, in describing the human body, commences with the bony foundations, and then goes on to the muscles, viscera, &c.

If there be any exception to this steady progress, it is to be found in the momentous transition from the concrete to the abstract, from the particular to the general. It is partly true that by the preparation of particulars we advance to generals, but there is not the same imperceptible transition, or unbroken continuity, as in proceeding from syllables to words, from a tree to a wood, from an easy air to one slightly more difficult. There is a certain jump in passing out of the life in particulars to the life in generalities ; we feel ourselves taken up into a new sphere ; we are called upon to exercise a new kind of faculty.

For this novel effect there must be a distinct phase of brain-development, and, therefore, a certain age attained, irrespective of the amount of preparatory impressions. The law of logical sequence merely includes the fact, that the concrete must precede and the abstract follow : there is, however, much else to be considered. Seeing that a vast compass of educational method and procedure is regulated by the transition, its minute conditions and circumstances need to be unfolded once for all.

We may and do proceed with the *classing* operation, from the very first, and without break or interruption. The child discriminates and identifies ; when it has identified a number of things of the same kind—chairs, spoons, fires, dogs, human beings—it has formed classes ; it has attained generality together with particularity. Yet these classings do not amount to Abstractions.

They do not proceed far enough to bring out the difficulties of the generalizing operations.   Many classes are
formed that are but one stage removed from the particulars : water, food, toys, lights, trees, horses ; over all the
region of experience these short leaps are made from
an early period ; while there is little progress anywhere
towards the higher flights.

It is for such ulterior flights that a certain maturity
of mind is needful ; a particular moment of advancing
strength, when the conceptions can be carried to the second and higher degrees of generality.   This is the stage
when we must be prepared to handle symbols, to pass
from sense perceptions to abstract conceptions ; when
we can manipulate numbers and forms, having no apparent reference to particulars at all.

Without much prompting, the child goes on accumulating classes of the first degree, and would go on to
the end of life in the same course.   It is only by express teaching that it climbs to the higher degrees— to
take cognizance of a piece of furniture, a tool, a quad-
ruped, a sum, a sensation, a society ; and a very large
part of teaching is occupied with this work.   It comes
up in season and out of season ; and the teacher's
resources should always be equal to it ; and at any rate
he should know whether or not it is within his compe-
tence at the time.   He cannot be too well informed as to
the conditions of success in explaining and impressing a
generality.   Indeed, this is the central fact or essence of
Exposition, properly so called.

It is, I repeat, universally admitted that, for a General
or Abstract Notion, the essential preparation is the Particulars.   But a great deal has to be taken into account

besides this obvious fact. The mere presence of the particulars does not suffice to evoke the generality. The number and the character of these must also be taken into the account: they may be too few, or they may be too many; they may even have the effect of obstructing the growth of the general idea.

1. In regard to the Selection of Particulars. This must be such as to show all the extreme varieties. Identical instances are not to be accumulated; they merely burden the mind: varying instances are necessary to show the quality under every combination. To bring home the abstract property of Roundness, or the circle, we must present concrete examples in varying size, colour, material, situation, and circumstances. To explain a Building, we must cite instances of buildings for all kinds of uses.

The best instances to begin with are those that show the main idea in prominence, and the adjuncts in abeyance. We cannot command a circle in the abstract, as Plato imagined ; and we cannot present one in the concrete without size ; but we can reduce the material to a thin black line on a white ground. Two or three such of different sizes, with one made of white on black ground, and one in some other colour, would eliminate everything but the single fact of form; this is as near to abstracting the property as the case admits of. If, on the other hand, we had proceeded from examples where the adjuncts are overpowering or interesting in themselves, attention would not have been gained to the form. The sun at noonday, the horizon viewed at sea, the circle at Stonehenge,—would be a very unsuitable selection for teaching the notion ; although after it is

otherwise gained, there is no difficulty in discerning it
in these instances.

2. The Placing of the instances should be such as to
bring out the agreements. If the objects are material
they should be similarly and symmetrically situated to
the eye. The comparison of numbers, as three, four,
five, should be in rows side by side to begin with. Cones
and pyramids are shown to the eye resting regularly on
their base. Vegetable and animal forms are symmetrically placed for comparison. This is the method required
alike for agreement and for difference.

In verbally described facts, the parallelism of the
forms of language is a well-known device of rhetoric.

3. The Accumulation should be continuous, until the
effect is produced. When we are bent upon driving
home a new generality, we should put everything else
aside for the time ; we should suffer no interruptions or
distraction. We are to accumulate instances of the
proper kind, and in the best order, until all disparities
are sunk beneath the pressure of the agreement. The
Theban Phalanx is the type of exposition for the general notion or abstract idea ; an overwhelming concentration at one point.

Many of our abstractions are gained by scattered
impressions, here a little and there a little. This is our
chance education, which, if least effective, is no doubt
least fatiguing. Whatever is gathered in this way is to
be accepted ; but the schoolmaster is not to repeat the
desultory circumstance in an express lesson. When he
enters upon an exposition, his business is to make it
continuous and thorough. If the pupils are ripe for
comprehending the notion of Inertness, a series of

examples should be arranged to make the general fact patent in spite of all disparities of accompanying circumstances.

It is the duty of the teacher to bring all the instances to bear upon the discovery of agreement. Any instance that is perplexing in itself will interrupt the general harmony. Still more will an instance that has a strong individual interest be an obstacle to the general impression. This is not sufficiently considered in exposition. Very interesting examples are sought with a view to engage the attention; they may succeed, but not in the way desired. Instead of leading the mind on to the abstract idea, they induce it to cling to themselves in their concrete or individual character.[1]

Contrast is an ever ready resource, and shortens the labour by excluding at once the notions liable to be confounded with what is meant. To drive home the idea of a circle we place beside it an oval. Along with groups of objects intended to give the abstract number four, we place a group of three and a group of five. White and Black are shown together. To explain more fully what luxury means, we adduce examples of simplicity and plainness of manners. The habit of assigning contrasts or opposites needs to co-exist in the mind

[1] In Macaulay's brilliant speeches on the Reform Bill, he urged, with unrivalled affluence of illustration, the growth and expansion of Great Britain, as necessitating changes in our institutions. His examples are numerous and telling, but also occasionally so gorgeous as to distract the mind from the general purpose by creating an interest for themselves. 'Who can say that a hundred years hence there may not be, on the shore of some desolate and silent bay of the Hebrides, another Liverpool, with its docks and warehouses and endless forests of masts? Who can say that the huge chimneys of another Manchester may not rise in the wilds of Connemara?'

of every instructor with the habit of quoting examples or particulars.

4. The natural inability to take interest in generalities, and the preference felt for the individual concrete, being the great obstructions to attaining general notions, we should clearly comprehend the counter forces in their favour. These are—first, the Flash of Agreement.

When among things that have formerly been regarded as distinct, there is a sudden flash of agreement, the mind is arrested and pleased; and the discovery makes one great element of intellectual interest, not only reconciling us to the general and the abstract, but, in the higher instances, imparting a positive charm. The disparity of the instances, and the previous labour of the mind in keeping hold of them, contribute to the elation of the discovery.

5. The other mode of overcoming the reluctance to pass from the interest of individuality to abstract notions, is the tracing of Cause and Effect in the world. The notion of cause and effect, the crowning notion of science, is one of the first to dawn upon the infant mind. The simplest movements are attended with discernible consequences: the fall of a chair with a noise; the taking of food with gratification. These instances are the beginnings of the knowledge of causes; and they are viewed correctly from the first. Now when any agent produces an apparent change or effect, it operates by only *one* of the many properties that it possesses as a concrete object. A chair has form to the eye, resistance to the hand, noise to the ear; and as these effects are seen in their separate workings, they

lead on to analysis or abstraction of the properties causing them. It is by the separation of effects that we come, in the first instance, to know weight as a property of things, and are able to regard the weight of a chair as a distinct property, owned in common with other objects. But for this experimenting on effects, we might not so soon or so readily depart from the collective individuality of things. The child knows what a seat in general is, by sitting first in its own chair and then on other chairs or on foot-stools ; by this experience it works its way to a notion of considerable generality.

6. For retaining a generality in the mind, the best way is to possess a good *representation* of particular instances. It depends on the character of the notion whether few or many are wanted. Very few are needed for a simple form—for weight, liquidity, transparency. For a metal, a plant, a tree, a bird, an article of food, a force, a society—a good many are wanted.

7. It is assumed throughout that the name is given along with the general notion ; to this, at the proper time, is added the Definition, which co-operates with the representative particulars in giving the mind a hold on the notion. The definition assigns some simpler notions, supposed to be already possessed; and it succeeds according as this supposition is correct. Thus then we have, as regards the circle, for example, (1) the representative instances, (2) the name ; (3) the words of Euclid's definition. The cluster so made up is our fully equipped notion of the circle.

In cases where a notion is formed out of other notions already grasped, the definition is a full and sufficing explanation, dispensing with its own particulars

This happens in Mathematics, when the mind is so well advanced as to be familiar with the elementary notions of number, equality, line, angle, plane, curve. It is a mere waste of time to dwell, at this stage, upon particular examples of triangle, square, polygon, circle, sphere.

In the ordinary course of miscellaneous teaching, a mixture of the two methods is unconsciously followed. A teacher in explaining 'kingdom' would both adduce individual kingdoms, as England, Germany, &c., and also define by language a kingdom as a people living to-. gether under one king. Most usually, perhaps, the definition is given first, and the particulars afterwards as examples. A river is defined—'a stream of water gathered by numerous rills from the high grounds and collected into one channel by which it flows to the sea.' The examples brought forward would explain the different parts of the definition.

So much for the great transition from the Concrete to the Abstract. The Analytical or Logical Sequence of teaching comprises the following distinct heads :—

First, and most obvious of all—from the Simple to the Complex.

Second—from the Particular to the General and Abstract.

These two must be regarded as fundamental, and almost exhaustive. There are, however, several important aspects of them that deserve to be signalized as if they were distinct cases.

Third—from the Indefinite to the Definite, or per- haps better, from the Unqualified to the Qualified.

We may be told a fact, in the first instance, in a vague, indefinite, unqualified form, as that all bodies fall to the ground : our next step is to learn it in its circumstantials and qualifications—the oblique descent of water in rivers, the rise of smoke, the belching up of volcanoes. The pupil in Astronomy is first told that the Sun is at rest in the centre of the system, while the planets move round it in circles. At a later stage, the circle is changed into an ellipse, with the sun in one of the foci. Then, the exact centre is shown to be the centre of gravity of the sun and all the planets. The withholding of important qualifications at the early stages is an accommodation to the pupil's capacity ; it can hardly be avoided, yet it needs management so as not to instil untruth.

Fourth—from the Empirical to the Rational or Scientific. This is really a mode of the transition from the Concrete to the Abstract ; yet it deserves an explicit consideration ; it marks with emphasis the arrival of the 'Age of Reason.'

Empirical knowledge is good and sufficient for many of the purposes of knowledge ; it may be all that is to be got on a given subject at any one time. Yet, seeing that much of our knowledge in the present day has attained the rank of scientific explanation, the pupil has ultimately to be put in possession of this higher form ; although, for a time, he may have to dwell in the lower region of the empirical. We first know day and night, summer and winter, the rise of the tides, the snow upon the tops of mountains, the falling of dew, the occurrence of storms, the dependence of vegetation on heat and on moisture,—as empirical. Our knowledge in this form may be on the whole very correct; our forefathers had nothing better to go by. And, in the empirical character, it suits an earlier stage of our education; we can understand a fact, as a fact, when we are incapable of comprehending the reason. Hence this is put down as one of the sequences or transitions in our progress. Nevertheless, it is essentially the transition from the concrete to the abstract : the reason of a thing is but a higher generality, into which it is resolved : the reason of the fall of bodies is universal gravitation ; of combustion, chemical union.

If anyone were to arrive at the maturity of intelligence, while still ignorant of a whole department of natural facts, such as Geology, the commencing with the empirical statement would be unnecessary. The concrete facts, in that case, might be given as scientific deductions from the laws, instead of being learnt in a provisional or passing shape. This method has its advantages. Many a one never knows the simplest properties of the triangle, parallelogram, or circle, until they are learnt in a course of geometry.

Fifth. In the culture of the power of Conceiving, the analytical order needs to be strictly followed. We must be familiar with the constituent colours and forms before we can conceive a new combination of them. We must know a marble surface and the cylindrical form, in order to conceive a marble cylinder. We must have seen numerous carriage-wheels, and have acquired the impression of gold, before we can conceive a gold wheel.

Sixth. We proceed from Outline to Details. This is the great maxim of the describing art, as in Geography. It applies to History also, although with a qualification.

Seventh. As a general rule, we proceed from the Corporeal to the Incorporeal, from the Physical to the Mental. The physical world is soonest understood ; yet from the very beginning we have some knowledge of the mental world; we learn to mark our own pleasurable and painful sensations, to enter into the pleasures and pains and passions of those about us. Our earliest literary interest supposes this power ; it is the basis of sensation narrative.

Such are the leading circumstances implied in logical or analytical sequence ; and if they could be followed strictly, the march of education would be clear. The reality is far otherwise. Many obstacles intervene, and it is well that we should be aware of these, that we may see how to evade or overcome them when possible.

To clear the way, we need to mark certain cases where Sequence does not necessarily apply.

1. The existence of *Correlatives* must be allowed for. Correlatives must be known together ; and although one is stated in advance of the other, the intended impression is not made till both are received. The strongest instance of this is the correlation of the Particulars and the General. One must be first, but both must concur before we have the meaning. The General is not understood without the Particulars ; the Particulars are nothing until they yield up the General. It is supposed that the Particulars must necessarily precede : this is not essential ; the generals may precede, and be held in suspense until the particulars are given. The order does not depend upon the fact

of correlation ; for neither term is intelligible without the other ; the meaning is a mutual embrace of two factors.   It is true that the mind must have a certain familiarity with concrete things before rising to the stage of generalities and abstractions ; but, at that stage, the things are not known as particulars fitting into a notion or a general law.   They are known in some different way, and have to be directed to the new purpose.   The child knows weights, but does not know them as examples of general gravity ; does not see through them to Newton's law of gravitation.   When this law comes to be taught, the particulars must be adduced on one side, and the general expression on the other, and, by the happy coalescence of the two, the law of gravity becomes apparent.   But whether the instances should be given first, or the formulæ first and the instances afterwards, is not always fixed.   It may be the shortest way of teaching in many subjects to give the generality first and the instances next ; the effect being suspended till both have come together. If the generality is not cumbrous or prolix ; if it does not involve a long series of abstract phrases, unilluminated by particulars ;—then, the best course might be to deposit it in the memory, for a little time, as an unmeaning formula, to be forthwith irradiated by the examples. In this way, the examples themselves are not kept waiting ; the generality is already there, and they have to fall in under it.

In the description of visible objects, Size, Form, and Colour must be given, but there is no natural priority. The mind usually waits till it learns them all. 'A black ball, a foot in diameter,' presents the three elements in the order—colour, form, size; which is as good as any other order, but not better.

In the various notions relating to Society, we have the same correlations, the same suspense of meaning till the correlations are adduced. A State implicates Ruler and Subject; neither is understood until both are explained. The order is immaterial.

In the Physical Sciences, also, mutuality of action is the rule. In the communication of force there are always two parties, the giver and the receiver, and one must be mentioned first; yet until the other is also mentioned, the fact is not complete.

Thus, then, provision must be made in the expository arts for bringing together correlatives in the way best suited to the several cases. The case is no real exception to the law of logical sequence.

2. The *mixing of notions of different degrees of advancement* and difficulty, is a thing that cannot always be avoided. An explanation should contain only matters already understood; but fully to adhere to this in the early stages is next to impossible. There must be mental blanks corresponding to many of the names presented to the young mind; sometimes, to a degree fatal to the understanding of what is brought forward; sometimes, permitting of a partial understanding, enough to be a stepping-stone to something farther, and in time to the complete knowledge of what is at present incompletely known. Although unavoidable, this is still an evil; and should be kept within the narrowest possible limits. Until all subjects can be composed on one level of intelligibility, and every subject have its proper order in the line of studies; and until no pupil be ever in-

trodnced to the higher without sufficient mastery of the lower, there will be these blank spots in the minds of learners ; the intellectual comprehension will be arrested every now and then from the want of some essential piece of knowledge.

In all subjects there must come up at the threshold names that cannot be adequately understood until the pupil has made some progress. A vague provisional surmise is the only thing possible ; perhaps the whole field of view is darkened in the meantime ; and yet it may be competent to go forward with simply glimpses of meaning. Partly by proceeding, in spite of defective insight, partly by going back to a fresh start, the various notions come to one another's aid ; what was dark grows clear.

It is in scientific or rational explanation, that is to say, science, properly so called, that the breach of sequence is most felt. When we are gathering in promiscuous facts, objects, impressions, without any attempt to explain, class, or reconcile them, we are not bound to any order. Whether we see a waterfall or a windmill first does not signify. So when our education consists chiefly in learning names, there is little that can be called sequence. Further, between one story and another story, one poem and another poem, there may be no priority assignable.

As a great part of early teaching is avowedly desultory, empirical, matter-of-fact, preparatory,—the order of presentation seems of little moment. The preference is determined by opportunity, and by the awakened interest of the pupils. Objects are impressed in the mind at the time when they are advantageously brought forward, whether in school or out of school. But the teacher should thoroughly understand the level that he is working at; he should not obtrude the connecting doctrines that make the knowledge scientific. The moment he aims at this, his situation is entirely altered; he must now chalk out a scientific scheme and follow it in rigid order.

Stories, poetry, histories, descriptions of travels, of places, of animals, are very mixed in their nature; the less intelligible and the more intelligible go side by side. The child picks up the crumbs of meaning and of interest that suit its advancement, and leaves the rest. There is no reason, however, why the effort should not be made to keep the whole composition to one level.

3. The *gratification of the feelings* interferes at this point, as at so many others. There may be enough in a composition to give pleasure, without its being understood. The mere form of poetry, the jingle of verse, has this effect: the melody of the words concurring. So there may be touches that bring out emotion, on a slender basis of understanding. Poetry serves this end; likewise the unction of religious and moral sentiments.

4. *Impatience to advance to matters of interest* may make us hurry over intermediate and preparatory matters, or proceed without these. This is an effect of solitary study; one of the uses of the public teacher is to stem the tendency. When left to ourselves we do not always see what necessarily lies between us and the goal we are aiming at.

5. The *language memory*, which is at its height somewhat prior to the maturity of the abstractive power, carries a great many things in the unmeaning state; and

the more we are endowed with it, the farther we can go
in dispensing with the full comprehension of what we
are laying up. It is often said that a knowledge of
meaning, in the shape of cause and effect, or other rational relationship, is the best aid to retentiveness; but
this is conditional. A good language memory dispenses
with all that.

We avail ourselves of the language stage of the
mind to forge adhesive links that are not so easy afterwards. Principles, maxims, theorems, formulas, definitions, that need to obtain a firm place in the memory,
may be given a little in advance of their being fully
understood. The licence must not be abused. For
one thing, the memory will not receive them, if they
are wholly devoid of interest; there must be something
either in the form of the words, or in the substance, to
engage the feelings, otherwise the anticipation is no
economy. Rules in verse have this advantage. A
scientific formula may have a certain pomp of language that impresses before it is understood. If the
subject affects the emotions, a faint glimmer of meaning
is enough; or one part understood may buoy up a good
deal that is not.

Pithy antithetic forms are easily committed in advance of the understanding of them. 'A line is length
without breadth,' is very abstruse in meaning, but very
easy to carry in the memory. 'All liquids seek their
level,' by dint of shortness and personification, ob-
tains an easy access to our stock of remembered forms.
The proverbial saws that we are accustomed to hear,
are stamped on the recollection long previously to our
being able to comprehend them. A long, prolix, un-

melodious, dry, and unintelligible statement might be committed through the urgency of the schoolmaster; but, however valuable it might be, in the day when it is fully revealed, there would be little gained by the process.

6. It is possible in the subjects that most depend on connection, *to pick up detached propositions* with their illustrations, and to hold them with a certain amount of understanding. This is, in fact, to repeat the empirical stage, after we are embarked in the rational or scientific career. A great many minds find themselves unable to keep up with the consecutive strain of a demonstrative science, and yet seize hold of portions of the reasoning, such as to pass muster in examinations; being only thrown out when called upon to reason from the commencement.

Like retaining knowledge by the mere language memory, this is a very insufficient mode of learning, and ends in the possession of scraps, without system or method, and without that reproducing power that a deductive science gives when once fully mastered.

7. It is not a breach of sequence *to cull precepts from different sciences*, and apply them to practice. The rules of Arithmetic can be put in operation without the reasons. This is still the empirical stage, where no sequence is observed. Provided only the terms of the rules are understood, we can carry them out in practice, while ignorant of the general subject, and wholly unable to give the reasons for them. In certain cases the working of the rules is not affected by ignorance of the sciences that they spring from; it is only in the higher arts, as Mechanics, Engineering, Medicine, Statecraft,

that the practitioner is to any great degree dependent upon a knowledge of principles and reasons.

8. The *cultivation of distinct organs* or faculties may proceed without any fixed sequence.   There is no settled order as between Colour, Form, and Number.   Singing neither precedes nor follows Drawing ; the culture of the singing voice scarcely depends on the articulate voice. The elegances of tone and manner in speaking can be given at any age : the only rule is to take the organs while plastic, and before they have contracted a wrong set.   So with carriage and deportment ; with dancing and gymnastic training ; with manual aptitude.   Again, moral training is not wholly dependent on knowledge and intelligence ; obedience and affection may take an independent start.   The morality that rests on reasons and consequences must wait till these are understood, and is much later.   Finally, there is no essential priority in the teaching of different languages ; some slight advantage is gained by taking Latin prior to the modern languages derived from it ; but as between German and Latin, there is no certain order.

The power of Reading is not essential to information in things.   Knowledge may be communicated to any extent orally.   The proper time for beginning book acquirement is a matter for consideration and adjustment. Even long after a child can read, it is unable to extract much information from books.

9. The *knowledge of language* and *the knowledge of things* should proceed together.   Yet the pace of the two is not necessarily the same ; one may go faster than the other.   The knowledge of things growing out of personal and solitary observation does not carry

language with it. Communicated knowledge supposes language ; still the attention may not be equally directed to the facts and the expressions. Language may go on while knowledge is almost stationary ; that is to say, the pupil's facility in expressing the same things may be steadily improving. There is not a single fact but admits of half a dozen ways of being stated. Thus while language cannot be separated from things, it may be carried forward irrespective of any notable advances in knowledge ; and the same may be said of knowledge with regard to language. The mere *preference of attention* to one of the two members of a connected couple pushes forward the familiarity with that member, while the other is scarcely advancing at all. The effect is illustrated by the difference of the two kinds of minds —the language mind and the reality mind. A man cannot have the power of language, without things to apply it to ; but his fulness of expression may be out of proportion to his knowledge of the things expressed.

DOUBTFUL CASES OF SEQUENCE.

In the common routine of education, there is an order that may not be violated. The different stages of reading and of writing cannot be transposed. In all the mechanical arts, certain simple movements have to be mastered, and are then conjoined in more complicated operations. There may occasionally be a little doubt as to which of several movements should be first ; it may want a subtle analysis to decide which is the most elementary of. two acts ; as whether straight strokes or pot-hooks are the best commencing exercise in writing.

In regard to Arithmetic, the only question is the

order of the Empirical and the Rational stages. In the newer methods of teaching, what used to be purely empirical, mechanical, or memory by rote—as the addition sums and the multiplication table—is now made to a certain extent rational from the beginning. That is to say, by concrete examples, it is shown how 4 and 5 make 9, and 3 times 7, 21 ; and on the basis of the concrete illustrations, the equivalent sums and products are fixed in the memory. Still, this does not amount to Rational Arithmetic ; it goes a little way, but not far. Children can with difficulty rationalize vulgar and decimal fractions ; and hardly at all the rule of three. The memory for the Tables, and for the manipulating of fractions, advances much faster than the comprehension of the reasons ; and it is not desirable to face these at the age when they are not readily intelligible. There is plenty of interest in the operations without the comprehending of the scheme of mathematical demonstration ; the ability to work out the prescribed exercises brings its own reward.

In certain respects this knowledge is highly scientific ; the terms are all clearly conceived, the directions precisely followed, and the results accurately arrived at. There is nothing slipshod, no vagueness to be corrected, nothing to be unlearned. The theory, rationale, or demonstrative connection of the steps is alone wanting ; and that is a later acquirement.

There is no exact parallelism between Arithmetic and Grammar, and no motive for priority, one over the other. Grammar has only in a very vague form the division into two stages, each perfect after its own kind. The order of grammatical teaching given in the

Standards of our 'Code,' does not represent the real sequence of study. In the Examination in Standard II., 'the scholar is to point out the nouns in the passage read ;' in Standard III., have to be pointed out 'the nouns, adjectives, and adverbs ;' in Standard IV., are required the parts of speech at large; in Standard V., elementary analysis of simple sentences ; in Standard VI., grammatical analysis in general.

The spreading of the Parts of Speech over three years appears a most arbitrary proceeding. The assumptions underlying it—namely, that the child can comprehend the Noun a year sooner than it can the Adjective and Verb, and these a year before the Pronoun, Preposition, and Conjunction—are not based on any facts or reasons. The Pronoun can be taught as soon as the Noun is understood ; and the Verb, Adverb, and Preposition are all linked together. Moreover, if the Parts of Speech are to be properly taught, the Analysis of Sentences should come forward at the very beginning.

Again, as the Parts of Speech must be all understood before any grammatical rule can be given, or any error be corrected on grammatical principles, there is, on the above plan, an enormous suspension of the practical interest of the subject. For two years, at least, all is barren. This circumstance alone is a great waste of power. In Arithmetic, the fruits of the teaching are reaped almost from the commencement ; questions are worked, and applications made such as the pupil can feel humanly interested in.[1]

---

[1] English Grammar has been passing through a revolution in the course of the last thirty years ; the definitions of the Parts of Speech have been vitally changed. I have dwelt upon this subject in another place (see

As regards sequence in learning the mother tongue, there is no particular order in the vocables themselves: these arise with the occasions, and with the subject-matter or things. The grammatical or structural part is learned at first by hearing and repeating connected sentences. If those are obtained from correct models, the child learns correct speaking at once; and may master all but the most delicate refinements of speech without any grammar or scholastic teaching. Everything is learnt empirically; reasons are neither given nor sought, being for the most part unnecessary.

The ordinary English teacher has to deal with pupils in every way deficient as regards the power of speaking English. Not to mention that their information is in a backward state, and with that the language suitable for conveying it, pupils express what they know badly; they have not either number or choice of words, they have not a command of sentence arrangements; their forms of expression are positively bad, whether as grammar or as idiom. If they could be made to understand the grammatical science, that might be the shortest way to their improvement. But the work of education is commenced long before this is possible, and they must be instructed empirically, the reasons being delayed for several years. It is absurd to suppose that the know-

Companion to the Higher Grammar). The old definitions rendered the science profoundly illogical, and but little suited to its main purpose.

I do not believe that Grammar in any shape can be a scientific or logical discipline. I doubt if there be any practical science (except Logic itself) that can impart a training beyond its own purpose; and Grammar is not an exception, although attempts are made to render it such. I have given the reasons very fully in an article on 'Teaching English' in the *Fortnightly Review*, for August, 1869.

ledge (if such it could be called) of three out of the seven parts of speech could make a basis for approaching the scientific explanation of good grammar. In the reading exercises, in the teacher's spoken address, the pupils hear proper and correct language, as well as choice and effective language. In their own answers, they perpetually fall short both in grammar and in other merits; and they are to be corrected on the occasion, and told what they ought to say; reasons being as yet withheld. Their known provincialisms are expressly called out in order to be extirpated. Even if they leave school before the age of grammar (which I think is not earlier than ten or eleven) they should still be sufficiently disciplined in correct speaking to rise above the prevailing vulgar errors, if not to attain a better style of speaking and writing in the whole.

While there is no natural priority as between the two subjects of Rational Arithmetic and Grammar (which is rational from its very nature), of the two we may pronounce Grammar much the hardest, and requiring a riper state of the faculties. In point of difficulty, I would compare Grammar to the commencement of Algebra; meaning by Grammar—Analysis of Sentences, the Definitions of the Parts of Speech, and the equivalent functions of the single word, the phrase, and the clause. There are easier parts of grammar: both Inflexion and Derivation are easier than Parts of Speech and Syntax; but it is scarcely worth while introducing these much before the age when every part of Grammar may be understood.

There is abundant occupation from six to nine years in knowing words and in getting sentence forms impressed on the memory, together with pronunciation and good reading. In addition to committing poetry, short pieces of prose may be selected for the goodness of the style, and also committed. There is no need ever to take a sentence to pieces. Sentences may be varied to show the same thing differently worded; and the pupils will gradually feel the superiority of one form over others; while, as regards the correct and conventional idioms, they must take the merits upon trust. (See Chap. IX.)

Still greater difficulties of sequence are found in the *knowledge* communicated during the first few years of school training. The composition of the Reading-books shows the prevailing views as to what subjects are to precede and what to follow one another. There is usually a mixture of easy poetry, tales, mostly with a moral, and simple information on interesting subjects within the capacity of children. The ends sought are to give pleasure, to cultivate the affections and the moral feelings, and to make a beginning in the imparting of knowledge, or rather to follow up the desultory impressions of personal experience, with connected statements that shall extend, rectify, and concentrate these chance impressions. Nevertheless, exercises in spelling, pronouncing, reading, and knowledge of language, at first take the lead.

For interest or amusement, the tale or narrative is the central device; and the art of weaving suitable tales has attained great perfection. A piece of information, a moral lesson, can be wrapped up in a short tale, and brought home with impetus. As there is a considerable expenditure of mind in proportion to the result, the information or moral should be well selected ; every little point in the vast area of useful knowledge cannot afford the requisite machinery.

Next to the tale (which may be prose or verse) is the poem, or metrical composition. The special advantage is the impression made on the ear, and through that on the memory; there is also the loftier strain of diction, which the child is gradually led up to appreciate. Poetry is used for moral lessons, and also for condensing information : the months of the year, the characters of the seasons, the habits of animals, the description of flowers, the events of history, are embodied in verse, for better lodgment in the memory ; being more agreeable as communicated in this form. The most amusing of all is the bold imaginative fiction, wrought up even to extravagance ; this can barely be allowed to pass as culture, although that claim is sometimes made for it under the head of Imagination.

There is no real question of priority, until we look closely to the kind of information brought forward in the separate stages of the reading lessons. On this matter, teachers seem hitherto to be only feeling their way ; it is no easy task to chalk out a course that shall be really consecutive. For one thing, it is difficult to gain an adequate view of how much knowledge the child of six or seven brings with it to found upon. The experiment is not easy to make, owing to the very erratic character of a child's promiscuous impressions. What is still more serious is the difficulty of laying hold of anything in the nature of information that is worth communicating, and yet does not shoot too high.

If the early training could be so directed as to en-

16

rich and invigorate the conceptive faculty, a time would come when definite knowledge could be absorbed so rapidly as to dispense with the attempts to impart it prematurely. All the information in a reading book for the Third Standard, which is spread over a year, could be taken in by an apt boy of fifteen in three weeks.

With an eye to such training, it is obvious that the memory for outward things, for sights, sounds, movements, must be carefully nurtured. A certain range of objects must be duly impressed in the first instance, and something done to cultivate new constructions or combinations of them.

The regular course of school training has this effect, if it has any effect at all ; but it is under the guise of contributing definite and finished information about this, that, and the other thing. The so-called culture of the Imagination is, in the first instance, the stirring of emotions pleasing to the young. By virtue of the emotional excitement, certain pictures, images, or descriptions are impressed ; and these are a part of the permanent conceptions of the mind, useful on their own account, and also as materials for working up other conceptions. The more extravagant they are, the more they contribute to present emotion, and the less to the stock of useful conceptions. ' Jack and the Bean-stalk,' ' Cinderella,' ' Puss in Boots,' *et hoc genus omne*, possess very little cultivating power. It is to stirring incidents from real life that the advantage in this respect most decidedly belongs.

In the present connection, we may take a further glance at the Object lesson and the manner of conduct ing it. The teacher can make anything he pleases out of this; it may aid the conceiving faculty, or it may not. The first good effect of it is, to waken up observation to things within the pupils' ken; by asking such questions as will send them back to re-examine what they have been in the habit of slurring over; or by questioning them on objects actually present. This is the beginning of the culture proposed. A wrong direction is given to the lesson when we assume the pupils' capability of bringing to mind whatever things they may have once seen, and when we expect them to make these up into new combinations.

The basis of the conceptive faculty is necessarily experience of things; of scenes, human dwellings, inhabited cities, and all their component parts, living beings — men, animals, plants—operations and activities, social gatherings and intercourse. The wider this experience, the better is the commencement. Next to experience are motives to attention or observation; these belong to the character of the mind, and cannot be artificially produced, except to a small extent. Intelligent companions are the best fostering causes of the requisite attention. We cannot secure strong emotions at every point; and even if we could, a more moderate excitement would be better in the long run.

The teacher might try to realize the situation of the child in its random accumulation of experiences, and to play up to it. Opportunities arise for stimulating interrogatories such as may quicken the retrospective glances at what has been experienced, and sharpen and point the attention for the succeeding opportunity. Books can hardly be contrived so as to hit the mark

Indeed, it is not easy to conduct a class in this line of exercises, wherein individuals differ so much. The routine of teaching as prescribed for teachers generally cannot readily be accommodated to the purpose; although the end may be brought before the mind of every instructor.

Let me advert still further to the composition of the ordinary Reading Books, as regards the point of sequence. The subjects of the first standards are simple poetry, fables, anecdotes of animals, easy stories. This is the stage when learning to read is the chief end of the lesson, the subjects being immaterial and secondary. In so far as the child reaches to a comprehension of the meaning, it finds gratification for the simpler affections and emotions, and is not expected to appropriate much information. Moral lessons are never lost sight of. Biographies of good and eminent persons are an early meal.

In the second and third standards, while the poetry is more varied, and the stories more lengthened, definite information begins in various forms. Natural History turns up one large region that is persistently drawn upon. Next come the two wide-ranging departments—Geography and Civil History. A further branch is Useful Knowledge—respecting the arts, industry and usages of life. All of these are given at first on the desultory and empirical plan; and, not till the higher Standards have to be used, is an attempt made to be more consecutive. Elementary notions are afforded of Physics, Chemistry, and Physiology; to these belong a rigorous sequence, unless we still adhere to the desultory and empirical treatment, which is, properly speaking, not science, but the preparation for it.

Natural History has three principal divisions, corresponding to the three kingdoms—Mineral, Vegetable, Animal. A perfect understanding of these subjects, (not including a mastery of all the details, but the knowledge of a certain number, with the ability to comprehend the rest when stated) supposes some knowledge of Mathematics, Physics, Chemistry, and Physiology. Minerals would come first, in the proper arrangement, then Plants, then Animals. But there is a kind of knowledge of the subject that inverts all this, and that is the knowledge afforded in the first Reading Books. Among the very earliest topics are the descriptions of particular species of animals, as the rook, the butterfly, the bee, the spider, the sheep, the camel, the elephant. The order of selection seems to be mere chance. The aim is to take up what is already in some way familiar and interesting, through actual acquaintance, or widespread repute. For example, every child has seen a butterfly ; could anything be better as the subject of a lesson ? On this slight basis of personal knowledge the reading book imparts a quantity of information belonging to the Natural History of the animal. It is of the class Insects ; its swallow-tailed wing is covered with a fine dust, which under the microscope is seen to be made up of scales ; it lives on the nectar of flowers, which it sucks up with a bony tube or trunk ; it has ten feelers or *antennæ* ; its eyes are composite or mosaic ; like other winged insects, it passes through various stages—egg, larva, chrysalis, butterfly. In the Third and Fourth Standards, this kind of lesson is frequent.

Again, as to Plants. These are by no means so popular; they want the interest of personality. Such wonders of the forest as the baobab and the banyan are brought forward and sketched so as to show their form, while description indicates their dimensions and other circumstances that inspire astonishment.   Flowers are ushered in to view, at first by the abundance of the poetry that embraces them, and in the incidents of gardening that occur in little domestic and other tales. Botanical knowledge comes much later.

Minerals are selected on similar considerations ; from their splendour, rarity, popularity, and other exciting circumstances.   There is, however, no conscious purpose in bringing them forward.   The author of a reading book seldom, so far as I have been able to judge, proposes to himself *representative* selections of the great departments of knowledge.

In these early reading books it is evident that the regard to sequence is very much dispensed with ; the supposition being that the stage for it has not yet arrived.   Nevertheless, it cannot be entirely dispensed with.   'A butterfly is of the class Insects ;' this refers back to some previous knowledge of insects as a class, and if no such knowledge has been given, the explanation halts.   The mode of putting it in that case should be different.   The appearance or features of a common butterfly would first be stated : these every child would partly recognize, while inwardly resolving to observe them better next time.   Then the less apparent organs might be noticed ; together with the microscopic additions.   This would be enough for the description at that stage.   Next the flight or motions might be stated. Then would come the mode of feeding, with which everyone can sympathize.   After which the marvels

of its transformations could be adduced in a general
way; all the better for the aid of diagrams or specimens.
Then, at the last, its belonging to the class Insect could
be so mentioned as to be a contribution to the child's
knowledge of the class; other familiar examples, as the
house-fly, the bee, the spider, being quoted. Even in
the desultory citation of interesting examples, sequence
is still an expository condition; the order of proceeding
from the known to the unknown, from the vague to the
precise, from the individual or species to the general,
must always be observed.

The interest of personality is perpetually inverting
the order of study. The child is introduced to men,
women, boys, girls, cats, dogs, horses, canaries, almost
before anything else. In scientific Zoology, the mental
qualities of animals are the last thing quoted; indeed,
these are often left out altogether. We acquire a certain
superficial knowledge of beings and animals, from attending to their outward aspects and movements, through
our sympathetic or other emotions. The Naturalist
begins at the other end; and it is no easy matter to
overtake him from our starting-point.

It is plain that we must work out three stages in
Natural History study. The first is allusive and desultory in the extreme degree. No order is observed
except to begin with what will afford interest to the
most juvenile feelings. It is a mere continuation of the
early impressions that animals, plants, and minerals
make on the mind by virtue of their chance interest.
There is, however, another stage, where there is provided
information of the scientific kind, only not in strict
scientific method. Here the order is far from indifferent

Whatever descriptions are given should proceed upon some prior knowledge, and should be kept in view, as a groundwork of something farther. The order of known to unknown, simple to complex, must underlie all teaching, however far it may be from the final, or third, stage of scientific order.

Let me next advert to the teaching of Geography, which is perhaps the most advanced in method of any subject, except Arithmetic. The sequence from the known to the unknown has been well worked out in the scheme of Geographical lessons ; the teachers in the German schools are very strong on this head. It is well recognized that the first notions of Geography are got from the child's own neighbourhood ; a hill, a valley, a stream, a field, a plain, must all be seen in the first instance ; and it is desirable that a plurality of each should come under observation before beginning Geography. Very young children cannot view these things in their geographical aspect. The ability to form pictures of mountains and rivers in other countries is a late stage of the conceptive faculty. The full notion of a river takes a great deal of thinking power, implicating as it does hill and valley, as well as the notion of an expanse of country made up of these, and constructed so as to converge to one main channel. All these elements have to be dealt with in separation, and yet in a well-considered order, as so many object lessons.

Concurrently with this effort is the mastering of direction, and the cardinal points. This is one of the earliest abstractions that the pupil is expected to master.

being coeval with the higher stages of Arithmetic. It can be very successfully conducted upon the immediate surroundings of the school, and these can be put into their Geographical relations at the same time; while the imagination can be conducted to the north and to the south, to the east and to the west, by naming more localities that stretch out in the several directions. The explanation of the four points can readily be carried up to the course of the sun, yielding at the same time the beginning of an Astronomy lesson, but the teacher should beware of pursuing these collateral lessons beyond his immediate purpose.

'The geography of the infant school,' says Currie, 'should be pictorial and descriptive. Commencing with the elements of natural scenery that fall under the child's observation, and carefully noting their distance and relative direction from the school, and from each other—the hill, the mountain, the brook, the river, the plain, the forest, the moor, the rich mould, the island, the sea, the cliff, the cape, the castle, the village, the city, that may be seen in prospect from the school ; the productions of his own land—its animals, its trees and flowers and herbs, its metals ; the men of his own land —their occupations, their customs, their habits, their food, their clothing ; it should seek to make the child realize the corresponding features of other lands and climes by comparison with what it has observed in its own. We should even set before his eye, when possible, specimens and pictures of foreign products and scenes, and for the rest appeal to his imagination to take off the impressions from our vivid description. Such is an outline in brief of the course the instruction should follow.'

It is difficult to believe in the possibility of such a course in the infant stage. It implies first that the child has had full opportunities of seeing places and objects. Next is assumed that the child's mode of looking at scenery has been elevated above its own petty amusements, and has seized the meaning of things in the great scale. Further, there is taken for granted the constructive or imaginative power of realizing other scenes differently arranged and made up. That any child before ten could be capable of such an effort is not to be credited. It begins to be practicable to the well-educated youth of twelve or thirteen, and approaches the greatest heights of a successful training of the conceptive powers.

Nevertheless, by a series of well-conducted object lessons, desultory to the superficial glance, but in the highest degree methodical underneath, the elementary facts of Geography may be gradually instilled, and a preparation made for the last stage of formal teaching by the maps. A great quantity of Natural History and other knowledge is taken for granted in carrying out the modern method of endeavouring to conceive in full concreteness the aspects of the various countries. But, indeed, it may be doubted whether so high an aim is really accomplished; yet, there is good done, and not harm, in entertaining it.

When the power of conceiving is sufficiently advanced, and when it has been fully exercised in geographical facts, the methodical study commences and is a tolerably plain course. The selection suitable for pupils at different stages and in different circumstances gives little trouble. The subject will come up again presently in connection with our next discussion.

Of all the departments of early teaching, none is so unmanageable as History. Its protean phases of information and of interest, its constant mixture of what attracts the youngest with what is intelligible only to the maturest minds, renders it especially troublesome in early teaching. Nothing comes sooner home to the child than narratives of human beings, their pursuits, their passions, their successes, and their disasters, their virtues and their vices, their rewards and their punishments, their enmities and their friendships, their failures and their triumphs. Arranged in circumstantial narrative, with the suspense of a plot, and the sensational conclusion, these incidents of humanity arouse our feelings and interest, at the first dawn of intelligence, and never lose their magic.

Narratives, as we have seen, come on the stage to lighten the toil of learning to read ; they are not further counted on, except for making an amiable or moral impression. Gradually they are made the vehicle of easy kinds of useful information, but are not yet thought of as supplying historical knowledge. In the biographical form, they begin to enlarge the acquaintance with human beings of the more eminent class, but with a view to excite emotions in the first instance. The narratives of social collective action, which alone is properly historical knowledge, start with battles, which awaken the early and powerful passions of the mind, and give the first bias to the sentiments towards our own and other nations. The youthful mind soon comes to

understand the meaning of invasions, aggression, pil-
lage, conquest, on one hand, and victorious resistance,
on the other, with the incidents of co-operation and
alliances on either side.   In the course of these exciting narratives, there springs up a vague understanding
of the great fact of society—Sovereignty and Subjection ;
the parental sphere being a help to the conception.   By
degrees, the ordinary action of the sovereign power in
time of peace comes to be intelligible in the more prominent features, as administering Justice, raising Taxes,
and making Public Works.   With sovereignty attached
to one person arises the conception of the successive
reigns of the sovereigns, with which is associated the
mention of great events, and especially wars and other
explosive changes.

As with Geography, so with History, the first thing
is to familiarize the mind with the elements, or constituents of historical changes or events.   Only, these
are of a greater degree of complexity, and belong to a
far later stage.   Moreover, the child lives in the midst
of the simpler geographical elements ; views with its
own eyes, hills, valleys, plains, rivers, cities.   It is not so
easy to bring it into the presence of historical elements.
It knows family life, and a little beyond that ; it knows
of the policeman and his duties as representing in a
humble way the power of the State.   For historical conceptions, it must wait a much longer time, and take a
great deal upon trust.   But since the deep political
forces, which it cannot understand, take the form of a
stirring narrative, which it can in part understand, history
is seldom entirely devoid of interest or debarred from
leaving impressions, and in those impressions are mate-

rials that may one day constitute a portion of historical knowledge, in the highest forms. Children's history is simply the sensational events of history extracted with as little of the abstruser explanations as it is possible to give. It may be so conducted, and should be so conducted, as to impart an outline of correct chronology, which should be deposited in the memory at the earliest convenient moment when it is likely to be retained.

The reasons are obvious and many for beginning with our own country. We assume that there has preceded a view of the geography of the country, which fits into the history, so as to enhance the effect of both. Then all knowledge respecting the existing facts and arrangements of our nation—the Legislative, Administrative, and Judicial Systems, the Standing Army and Navy, the Religious Denominations in the three Kingdoms, Education, Agriculture, Trade, Manufactures,—assists in making intelligible the history of the past.

There can be no systematic teaching of History in school years ; but there may be an avoidance of perverse and erroneous methods. The attempt to plunge into modern European History at large with children of ten, can but confuse ; select episodes should be chosen on the ground of their impressiveness. The same in regard to Ancient History, with its more stirring incidents, and its gorgeous mythology, which, as being the creation of the infancy of the race, has power to arrest the infant mind in the individual, and is presented with this express view. Seeing that very little of real instruction can come of all this, the point is to see that it makes an impression on the feelings, and througn them on the conceptive power or the imagination ; if it falls flat, and has to be

inculcated by the force of discipline, it is better withheld.

In teaching Geography, slight touches of history may
be given, and in teaching History, geographical facts
may be impressed; due regard being had to the pre-
caution of not pursuing the digressions too far.

How to teach History proper, at the age when it can
be taught, resolves itself into the method of explaining
the elementary facts and workings of Government and
Society, or what is called Sociology. This might have
to be considered at the same time with the question of
introducing the laws of Political Economy, which form
a part of Sociology, in some respects simpler than the
laws of Politics at large, although in the end mixed up
with these. As repeatedly remarked, the stream of
stirring narrative carries with it a number of fragments
of a scheme of Sociology; and a time comes when
they may be pieced together and the scheme completed.

As History will always be brought into early teaching
long before the age when Sociology can be taught as a
science following on the Science of Mind, there must be
an empirical sociology involved or implied. This would
suit the middle period of a complete education, say between thirteen and sixteen, when by the present arrangements classical teaching is in the ascendant. At that time
the elements of Social Science might be introduced, and
might receive their illustration in the historical field;
but historical reading, apart from some definite social
conceptions, must still remain in the lower stage of
sensation narrative; or at most can but add to the
more ordinary facts of human nature.

The only remaining topic of Sequence is the order of the leading sciences—Mathematics, Physics, &c. If we take the five fundamental sciences—Mathematics, Physics, Chemistry, Biology, Psychology, the order now stated is what would be generally allowed. The Natural History Sciences—Mineralogy, Botany, Zoology, walk by the side of these ; Mineralogy, following on Physics and Chemistry, and Botany and Zoology being one aspect of Biology. Psychology, properly taught, would succeed Biology; but it is also the subject of an empirical teaching that dispenses with the knowledge and training of the preceding sciences. On Psychology would hang Scientific Sociology, fed by the earlier studies in Geography and History, but still demanding a rigorous scientific treatment in its place in the roll of the sciences. This would be the stage of Political Economy, and of the highest Ethical teaching ; but both of these are supposed to be previously given in the empirical shape.

CHAPTER VIII.

METHODS.

THE foregoing chapter is intended to relieve the present one by a separate handling of one leading topic of Method. A full consideration of order or sequence lightens the task now to be undertaken—namely, to set forth the methods of teaching in detail.

The Teaching Method is arrived at in various ways. One principal mode is experience of the work; this is the inductive or practical source. Another mode is deduction from the laws of the human mind; this is the deductive or theoretical source. The third and best mode is to combine the two; to rectify empirical teaching by principles, and to qualify deductions from principles by practical experience.

As Morals, Religion, and Art are not included in this chapter, the discussion will revolve on the one great topic of communicating knowledge, and will follow the various aspects that knowledge assumes—as particular or general, and as relating to one or other of the many departments of the knowable; for example, the various sciences, in so far as they differ in their methods of teaching.

The arts and devices for communicating knowledge are comprised in the practical science of Rhetoric, and ought to be exhaustively viewed in that science.   Rhetoric, however, has not yet been so completely shaped as to supply everything that belongs to the various emergencies of teaching.   Nevertheless, the study of that subject, so far as it has been matured, is in the direct line of the teacher's work.   The practice of the school not being confined to the means of assisting the understanding, but involving also appeals to the feelings, all the parts of Rhetorical method may come into operation.

Still, Rhetoric, as usually given, leaves out many points relative to the work of the school.   The Rhetorical arts of good exposition, by Example, by Contrast, by Illustration, by Proof, must be known to every successful teacher ; but the ordering of lessons, the conducting of *vivâ voce* interrogations, the proportioning of oral instruction to book work, the managing of object lessons,—demand an amount of consideration that they have never yet received from any writer on Rhetoric.

The outline formerly given of the great functions making up Intellect, supplies the leading points of method, as regards knowledge generally.   We have seen what arrangements favour Discrimination as such ; and Discrimination is not only the beginning of all knowledge, but, under the more expressive form of the sense of Contrast, bears a part in every new acquirement.   The co-ordinate power of discerning Agreement has also its conditions, and these were previously stated, and again repeated in the last preceding chapter.   The great function of Retentiveness was likewise briefly unfolded, as to its manner of working, and the conditions assigned ; these being remarkably precise, as well as all-important.

In reviewing the various branches of school instruc-

17

tion, we can discern several common characteristics admitting of general treatment. In the commencement of Speaking, in Singing, in Writing, and in Drawing, we have mechanical constructiveness, and this has a still more extended application in the manual arts. The mode of working here is simple and uniform; its conditions have been already assigned [p. 40], and will now be more fully exemplified. In learning to Read, constructiveness is joined with the associating operation of uniting articulate sounds with visible symbols. There is also called into play the discriminating sensibility of the eye, on which depends the retentiveness or memory for visible forms.

Constructiveness, as distinct from literal memory, enters into all the higher education, and is described under various names, the most apt being Conception or the Conceptive power or faculty. The first foundation of this may be called memory, provided we understand that it is memory of the concrete, or the full sensible image of the things that have impressed the senses. Having been inside a great building, we carry away with us a more or less exact recollection of its form, dimensions, surface, and contents, in their order; this is memory, but it is also conception. The oftener we have been inside the building, and the more attentive we have been, the fuller and firmer is our mental image. To hold such recollections in our mind, is to conceive more or less perfectly what we have seen. This is a power and an education in itself; and it is the ground-work of the farther education of conceiving what we may not have seen, but merely hear or read about.

CONSTRUCTIVE ACQUIREMENTS.

We shall view these together, as they proceed according to the very same laws. The earliest acquisitions of infancy exemplify purely mechanical constructiveness. Such are Speaking, Writing and Drawing.

Reverting to the principles already laid down respecting the constructive process, we have first to lay stress upon the random or spontaneous commencement of our various movements. Action of some sort precedes the desired action; a great many movements are made before the proper one appears. The teacher cannot dictate the right movement; he must wait upon it, and try to clench it when it is at last hit upon.

Speaking.

The first lessons in speaking, gone through in the nursery, show the difficulties of commencement at their greatest. The school teacher finds the power in existence, and improves upon it. He has to impart new articulations and to correct and refine the old. He will encounter much stubborn inability to fall upon the desired sounds, and must proceed upon correct principles. His own articulation needs to be clear and expressive, for the sake of a good model. He must consider that this is one of the trying moments of instruction: all the circumstances need to be favourable; the pupils should be at their best, and in circumstances to favour vocal freshness and spontaneity. Many trials must be allowed to get a child into a new shade of vowel, as, for example, when Scotch children have to learn the English sound of 'all.'

Concurrently with the alphabet and the first lessons in reading, there is a great extension of the articulating range, to which apply all the maxims relative to every new constructive process. Time must be bestowed upon this part of the reading exercise by itself, irrespective of the farther operations of distinguishing and attaching the visible letters. The joining of syllables into words, is a matter of farther articulate constructiveness, and furnishes no small demands upon the flexibility of the articulating organs, as well as upon the cohesive or plastic power.

A good analysis of sounds, confirmed by teaching experience, shows the best order of the exercises in articulating. The vowels are indifferent in point of sequence; the consonants may show a gradation of facility. The combinations follow the sequence of simple and complex; but at every stage, it is a question of the compass and flexibility of the articulating organs, the beginnings being wholly at random. The teacher's opportunity is some chance hit, which he improves until the lucky movement is well confirmed.

This single branch of the reading lesson should have much time bestowed upon it. At the age when the communicating of knowledge is premature, the attention cannot be better occupied than with the mechanical accomplishments—of which articulation is at the head. The mere power of articulating should be followed up by elocution and cadence, which are equally suitable as subjects of training for the years from four to seven. To these also the same cautions are applicable; the concentrating of time and strength upon initial difficulties, and the patient waiting upon the pupil's own spontaneity, with the guidance of a clear model.

The Manual Constructiveness.

The school training in Writing and in Drawing is a branch of the training of the hand. The practice of putting children to write, as their very first attempt at delicate handiwork, appears objectionable. The art of writing ranks high among the manual acquirements, and should be preceded by easier exercises. The simpler lessons of Drawing are obviously easier than writing ; while the making of symmetrical shapes is more agreeable than forming letters. Probably the natural course to follow would be the method of the Kindergarten, which is to train the hand upon moulding objects in clay, followed by cutting out paper figures, and gradually leading up to elementary drawing, after which writing would come with comparative ease, but would still be a considerable step in advance, like beginning a trade.

The mechanical aptitudes have a Sense element, which must proceed with the active element. The child has to work up to some model or design, and must clearly perceive the appearances that it has to reproduce. This is described as the culture of the senses ; but it is rather the culture of the act or habit of attending to sense aspects and properties, and depends on evoking a special interest or aim. The interest may be the charm of the thing itself ; this may apply to little models to be imitated in clay, or to designs given for drawing, but cannot belong to alphabetical characters. There is also the interest of successful manipulation, which can be drawn upon after a little facility is gained. This belongs to the dullest subjects, and is the more needful in beginning to write.

While giving all credit to the course pursued in the Kindergarten, in devising preparatory manual exercises, before entering on the difficulties of writing, I must indicate what appears to be the danger and the abuse of that mode of proceeding. It seems a mistake to constitute these early exercises an end in themselves, and to allow the pupil to be absorbed and detained by them. A certain amount of manual power, and of sense discrimination, is necessary to everyone, for the various exigencies of life, and as a preparation for the higher knowledge; but it is only particular professions or trades that carry any one aptitude to high perfection, and the culture for a trade should not be set before the child beginning its education.   The drawing of symmetrical forms and elegant curves is a good thing by the way, as training the hand by something that possesses interest; yet is but a means to an end, and should be kept in strict subordination.   At some future day, select individuals will develope their forte or capacity for drawing, and render themselves skilled in it as artists or designers, only that is out of place at the commencement; and the indulgence of a special taste at the early stage but disturbs the proper career of the learner.

Apart from the systematic hand training of the Kindergarten, there can be no doubt of the advantage of combining writing with elementary drawing, as is advocated by Currie and others.   In the process of writing itself, the analysis of forms has to be taken into account, as is carried out in the method of Mulhaüser. This is merely exemplifying the order from the simple to the complex.   The only objection to the method is that it is dry and uninteresting; the pupil feels roused to a grander imitative effort by having a complete letter to form at a stroke. And although grown men, such as recruits in the army, can be kept at work on elementary movements, children have little heart in them, and are slow in conquering the difficulties that they present. It is like practising scales in music, which to very young pupils is repulsive. As soon, therefore, as the progress will allow, the attempt may be made to copy the complete letters, without surrendering the practice in strokes, pot-hooks, and the other simple elements.

The proper inclination, dimensions, and distances of the letters, are attained through a delicate sense of visible form which is very various in individuals, and is. best cultivated by drawing exercises. This need not be pushed to an extreme point of delicacy for the ends of primary education ; any very extraordinary endowment in the art is likely to be attended with deficiencies in other important mental qualities. All pupils should be brought up to the point of plain passable writing ; and should be made to put stress on the points that distinguish such letters as are apt to be confounded : it is not the schoolmaster's business to carry writing to the pitch of a work of art.

The Art of Drawing, here invoked as a coadjutor of the first steps in writing, is, I presume, sufficiently well formulated on its own account. I am interested only in the more elementary exercises, as I do not consider the higher stages to belong to general education. The exercise of perspective drawing from real objects, is thought a grand culture of the power of observation. I have already expressed doubts as to the truth of this view : the exercise certainly cultivates the observation of such points as are necessary for the purpose of drawing ; whatever is involved in these is attended to ; but obser- vation is a large word, meaning many things besides.

Reading.

The extent and complicacy of this accomplishment make it the work of years, even when not commenced very early. The power of speaking is presupposed, although it is in connection with reading that perfection in speech is ultimately attained. The eye and the in- tellectual processes bear the brunt of the acquisition.

The art of Reading should be viewed, in the first instance as distinct, both from spoken language and from the knowledge attained through speech ; it is also dis- tinct from the acquisition of farther knowledge through books, although intended to compass that object. It is the art of pronouncing words at sight of their visible characters.

If our language, like the Chinese, had a character for each word, the eye would have to be taught first to discri- minate the characters ; next an association would have to be formed between each spoken word and its charac- ter. The teacher shows the character and pronounces the word ; the pupil attends with the ear and with the eye, mostly with the eye, because the form is what is strange to him. We are not informed, so far as I am aware, of the methods of the Chinese schoolmasters for getting through the herculean task of forming several thousands of distinct associations between sounds and symbols. The experience of ages must have suggested the most economical mode, and it would be interesting to compare the approved method with what we should deduce from the laws of the Retentive faculty.

As an Alphabetical language, English is learned on the principle of analyzing words into their constituent sounds, and connecting these with the elementary or alphabetical letters. As an irregularly spelt language, there is still something of the Chinese necessity for taking each word by itself; we have to learn to pronounce 'rough' and 'through,' 'faculties' and 'facilities,' by looking at the words as wholes, and not by inferring from one to the other, or from the powers of the separate letters.

The first act of reading is to distinguish the letters by the eye, and especially those that are nearly alike. Here we fall back upon one main condition of the discriminative power—concentrated attention upon the difference; to secure which, we may magnify the difference artificially.

With the visible alphabetic characters or letters we must connect their names or vocal representatives, in order to speak about them, and with a view to the future stage of spelling.

The fixing of the visible impressions of the alphabet is hastened if the pupil is sufficiently advanced in the power of the hand to draw the letters with chalk, or with slate pencil. It need not take long to distinguish and name the characters.

Now commences the difficulty—how to deal with words. As these are made up of letters, it seems natural to jump from the sounds of the letters to the sounds of their combinations ; after knowing *p*, *u*, *t*, the child may be expected, on seeing ' put,' to pronounce it accordingly.

This might be the case, if the letters separately could be sounded exactly as they are in combination ; which is true of the vowels (allowance being made for our irregular spelling) but not of the consonants, as we cannot pronounce a consonant without a vowel, more especially the abrupter consonants, *p, t, k, b, d,* &c. The liquids, *l, m, n, r,* and the sibilants, are pronounceable without consonants ; but in giving them names, we still use a particular vowel, *em, ar, ess.* The pupil must be made aware as early as possible of this circumstance, by being initiated and practised in the effect of the consonants as they occur in words ; a thing that inevitably happens sooner or later, so that learners cease to be misled by the sounds used for merely naming the consonants. A little practice upon easy words, *pat, put, pop, tap,* gives a mastery of the value of *p* in composition.

Much stress is now laid by teachers on the point of beginning to pronounce short words at sight, without spelling them ; and a strong condemnation is uttered against the old spelling method. The difference between the methods is not very apparent to me ; after a few preliminary steps, the two must come to the same thing. Of far more serious import is the mode of grappling with irregularity of spelling. When among the earliest lessons, a child is made to pronounce—'do I go—is it set on,' it is on the Chinese principle of learning each word *seriatim*, without inferring from one to another ; the *o* is sounded in three ways, the *i* in two, the *s* in two. After a time, no doubt, the letters are found to have recurring meanings, and inferences from one to another may be made, with a certain allowance for two or three possible modes, the particular choice being decided by the word ; so that the Chinese principle is limited but not abandoned.

The preferable plan seems to be to carry the pupils forward a certain way on *perfectly uniform* spellings, so that they may get the idea of regularity, and also the most prevalent sounds of the letters. This is not so difficult upon *a, e, i, u*, whose short sounds, *at, bet, it, nut*, are almost uniformly spelt with a single consonant to follow. Moreover, the irregularities of the consonants could be kept out of sight for some time. Some notion of law and uniformity would be thus imparted at the outset.

For the long sounds of the four vowels, there are usually employed some additional letters, unfortunately not in a regular way, but still serving as a contrast to the short sounds ; as *came, meet, sign, full*. These different devices should be classified, giving the most frequent first, and then the less frequent.

The refractory vowel is *o*. If when our language became possessed of the sound *awe, all*, a vowel character had been invented for it, we should have been saved a large number of our worst spelling anomalies ; if that could be done yet, it would be our greatest phonetic improvement.

The modes of spelling for this sound may still be classified, but they are numerous and contradictory—*all, fall, call* (*cf.* mall, sh ll), *cause* (*cf.* aunt), *awe, talk*. With the short sound of this vowel, *got, not, rot*, the system of the other vowels prevails, but with exceptions—as *God, Job, both, loth*. Still, uniformity should first be taught, and the exceptions enumerated.

The real difficulties of our spelling are nearly ex-hausted upon our monosyllables; if these were fully mastered, the anomalies in words of more than one syllable would not seem formidable.

Notwithstanding the zeal that has been displayed in the work of phonetic reform, no one seems to have gone through the labour (not small) of classifying the existing spellings under uniformities and exceptions; proceeding upon such classings as give the most agreements and the fewest exceptions. Until this is done, learning to read is not made so easy as it might be made. The principle of minimizing exceptions, and of placing them all together at the end of the rule, is the only known principle of economizing the learner's strength, or of reducing the Chinese operation to the narrowest limits.

After the very best classification, the attainment of English spelling is a work of long time and detail, the result of combined reading, writing to dictation, and extensive practice under correction.

Pronunciation follows in the same course, and is usually connected with reading. It can be taught only by teachers that themselves pronounce well. It is conducted on the plan of attacking the prevailing errors and faults of the children, which are for the most part local or provincial. A phonetic spelling would be a valuable help to pronunciation.

Good elocution is a still higher aim, and must come later, as it supposes that the pupils are alive to the meaning of what is read or pronounced.

Division of labour requires that the attention should be concentrated on the act of learning to read, without endeavouring to extend the bounds of the pupil's knowledge in the first instance. The reading exercises must refer to some subject or other ; but the proper plan is to take very familiar and easy subjects. Indeed, the subject matter should excite as little attention as possible, and the visible words as much as possible. If the mind is to be in anywise occupied with the meaning, amusement should be the aim, by way of relieving the strain. Some of the emotions may be occasionally touched—affection, power, admiration, indignation ; and it is so far well that these should have a good moral tendency ; but even moral teaching, if fatiguing, is to be foreborne. The little lessons about cats, and dogs, children at play, and kindness to those in distress, are intended to give scope to the emotions of children—more particularly the agreeable patronizing emotion—by suitable stories and situations ; and this is the reward for the fatigues of commencing to read. In themselves, these themes go for next to nothing. Even the pretty little poems are of so childish a character, that it is better they should not be remembered at all, unless as part of the stores of the future parent.

During the first year or more of learning to read, the extension of knowledge should still depend partly upon personal experience and partly upon oral communication. There comes a time, however, when the book read is regarded not merely as an instrument of instruction in reading, but as a vehicle for information. This is a critical moment, a new start, although usually disguised by the stealthy way that it is brought in. The situation is one that needs to be carefully considered, and the conditions of success fully understood.

Already, in discussing Sequence, I have alluded to the nature of the progressive lessons in general know-ledge and to the difficulties attending it.   We may here narrow the issue, by considering what things to avoid as unsuitable, or else unnecessary.

Assuming as granted, that we should not enter upon matters either beyond the comprehension of the pupils, or beyond their interest at the time, the teacher should avoid interfering with their own spontaneous course of self-instruction.   A parent can guide and direct this to a good result, but the means at the command of the teacher are much more limited.

The following is given by Mr. Morrison, as an example of a lesson on the simplest conceivable subject, used in the first instance for practice in reading :—

'The rat sat on a mat, the fat cat ran to the mat, the rat ran in-to the box. Can the cat go in-to the box ? no, the fat cat can-not go in-to the box.'

Now this lesson is contrived purely with a view to words and spelling, and although the words are put together to make a meaning, the choice is guided solely by the aim of exemplifying certain vowel sounds.   The early introduction of the 'cat' and the 'rat' to the notice of children is due to their being convenient examples of monosyllables in short *a*.   The 'mouse,' the more usual object of the cat's activity, is kept back because it is a more difficult spelling.   Now, it must be allowed that the relations of the cat to the rat do possess a natural interest of a kind to affect the juvenile mind.   Predatory pursuit excites us from the earliest years ; and any incidents embodying it will waken up the feelings and exercise the imagination in a bloodthirsty chase ; thus enlivening the dull and dreary exercise of learning to read and spell. It does not follow, however, that the subject should be drawn out as a lesson in useful knowledge, by turning it round and round, by making new suppositions as to the relations of cat and rat, and asking the pupils to say what would happen under these altered relations. There will come a time, and a place, for this sort of exercise, but the choice of subject should then be governed by its drift or meaning, and not by the words that happen to clothe that meaning. The following is the line of examination suggested by Mr. Morrison :—

What two animals does your lesson speak about? Have you ever seen a rat? A cat? Which is larger? Which is stronger? Where was the rat sitting? What was it doing on the mat? What was sitting on the mat? What is a mat? Where do you see it? What is its use? If a little boy get his shoes dirtied, what should he do before going into the house? The mat is used for—*wiping the shoes*. The rat sat on—*a mat*. Was that its own place? Where should it have been? As it was sitting on the mat who saw it? What kind of cat was it? And what did the fat cat do? The fat cat ran—*to the rat*. (Describe the running—show how the cat would sit and watch, and then bound forward. This will amuse and interest the children, and keep them fresh for the remainder of the examination.) Do you think the rat would wait on the mat? What would it do? It would—*run away*, run away to—*its hole*. Where did it run? What is a box? What made of? How would it get into the box? What must have been in the box? You see then the rat ran into—*the box*, through—*a hole*. Did the cat go into the box? Why not? The hole would not let in—*the cat*, but it let in—*the rat*. Would the cat go away from the box? What would it do? It would—*watch*, beside the—*box*, to see if the rat—*would come out*, &c.

Some of the criticisms suggested by this line of

questioning belong to a later discussion on the Object Lesson. At present we remark that if the intention is to base the examination on the child's experience, the cat and the mouse would be more suitable, as more likely to be witnessed. The child has little opportunity of closely inspecting a rat; and even the play with a mouse is one of the rarest treats of the child's experience. The cat with her kittens would give a firmer basis of the actual; and might comprise the higher situation of her jealousy of the dog's attentions.

But the main point to be insisted on at this stage is, that, while it is right to compose little scenes, situations and actions, to relieve the dryness of reading exercises, these are not necessarily suited for cross-examination, with a view to extend the knowledge, or to sharpen the faculties of pupils. Any meaning that may attach to the compositions used for learning to read, serves its purpose if it slightly amuses and interests the child; if it deposits a moral or a fact, so much the better, but this should not be insisted on, nor should the teacher consider it his duty at this stage to impress the meaning. When he comes to that part of his work, he must have compositions expressly suited for the purpose, and not shaped for another purpose in addition. No man can serve two masters; scarcely any composition lends itself equally to teaching language and teaching knowledge.

I do not maintain that the attempt to improve the knowledge and intelligence of children should be postponed till they are good readers; but I hold that the exercises should be disjoined, and grounded on different texts. The same text *may* be used for both purposes, but it is too much to expect that what is best suited for language, should be also best suited for meaning.   Nor should the lessons be intermingled ; times should be set for each.   There may be many examples of 'good thoughts well expressed,' but it is not likely that the expression and the thought shall both fall in at the same stage of the pupil's progress.

The discussion of Method, if taken in the usual course, would lead us next to Arithmetic, Grammar, Geography, History, among elementary studies; for the higher studies, Languages (foreign) and Sciences.   It is desirable, however, to consider, with some degree of closeness, the Object Lesson, which is the precursor to the more systematic handling of the various branches of natural knowledge, and, from its undefined character, is more apt to run in unprofitable channels.   The Object Lesson is exemplified in the Standard Reading Books, and the teacher may strictly follow what is there pro-vided for him ; but he is also directed to give such lessons from his own invention.

THE OBJECT LESSON.

The Object Lesson is made to range over all the utilities of life, and all the processes of nature.   It begins upon things familiar to the pupils, and enlarges the conceptions of these, by filling in unnoticed qualities. It proceeds to things that have to be learnt even in their primary aspect by description or diagram ; and ends with the more abstruse operations of natural forces.

The dangers attending it are :—(1). Superfluous communication, or the occupying of time with what the children quite well know, or will soon know of their own

18

accord, by observation and interchange of thoughts with
parents and companions. (2). Assuming what is not as
yet intelligible to the pupils, or not sufficiently so to
be a stepping-stone to higher knowledge: an error apt
to be committed in every stage of teaching. (3). Un-
seasonable and uncontrolled digressions; this evil we
shall have to put in the strongest light. (4). Absence of
consecutive arrangement; and, generally, of instructive
relationships, and mutual lights.

Antecedent to all considerations of choice, arrangement, and handling of such lessons—are the fundamental
laws of explanation by Agreement and Contrast, the laws
of the Abstract Idea, and the course from the Known to
the Unknown, the Simple to the Complex, the Empirical
to the Rational. The mind of every intellectual instructor needs to be rooted and grounded in all these matters,
so that they may become omnipresent in the details of
teaching.

To lay down the proprieties of the Object Lesson,
we must endeavour first to classify its different forms,
and to ascertain its exact purpose under each. An
order or sequence is assigned corresponding to the age
of pupils, and this order supposes that the kinds should
be well classed.

Pestalozzi, one of the first propounders of the Object
Lesson, regarded it merely as the proper way of teaching the use of Language, that is to say, it provided the
means of knowing the things expressed by words. But
knowledge has a prior and independent value, and is
not an incident of correct speaking; and we must look
at the lesson simply as a mode of imparting knowledge.

The Object Lesson passes by Arithmetic or Number,

the exercises in Form and Colour, Geography and History. It introduces the pupil to three great fields—
Natural History, Physical Science, and the Useful Arts,
or common Utilities of every-day life. The most usual
direction for conducting it is, first to point out the appearance or sensible qualities of an object, and next
to specify its uses. A better rule would be, to give the
uses first (after the more obvious aspects); use is quality
in act, and our interest in things is first excited by their
active agency. Take, as an example, a piece of glass.
This is held up to the view of the pupils. They have
already had occasion to see and handle glass; they
know it in windows, in table glasses, in bottles, in
looking-glasses and ornaments. It is purely a thing of
use; it is brought into existence for use. What, then,
should a teacher say about it? He need not tell that it
is hard, smooth, and allows things to be seen through it;
all this the pupils know. They also know that if it is
struck or falls, it is broken, cracked or splintered; and,
further, that splinters cut the hands very readily. As
an exercise of sense perception, there seems nothing to
be added to the knowledge of any child of five or six on
the ordinary properties of glass. The teacher may get
into a conversation with them, and make them *express*
their knowledge in words, so as to show that they have
been observant, and farther that they have names for
embodying and communicating their experience. This
much may be valuable as a stimulus to observe, and as
an exercise in language.

The perplexity begins, when it is proposed to extend
this sense knowledge, by the recital of unobvious or
hidden properties. The teacher has now twenty outlets,

and which shall he choose? Is it to be the uses that lie outside the scope of familiar observation? Is it to be the manufacture of glass,—including the materials that enter into it, and the various species of glass? Is it to be the discovery and history of glass? Is it to be the optical properties of glass? Is it to be the single property of transparency, illustrated by comparison with other substances? A teacher will no doubt feel at once, that to a particular set of pupils some of these things would be wholly unintelligible. There are, however, some points that would be within their capacity, and their interest; such as, the uses of glass beyond their own familiar circle, and perhaps the circumstances of its origin and history; also its component materials nakedly stated without the chemical laws of their union.

Still, even among the intelligible outgoings, there must be a ground of preference; some of them, it might be quite unprofitable to pursue at length. Uses that are mere repetition, or that inspire no interest, that might never be copied, that illustrate no important law of science—may be left without notice. The only point that readily occurs to me as worth dwelling upon is the leading circumstance in the manufacture of glass, the heating of sand in contact with soda or ashes. To pupils of seven or eight, enough might be said upon this point, to awaken interest, and to impress a fact for after use in teaching science. The striking changes produced by chemical combinations are highly sensational and can be firmly lodged in the memory, in set examples, before the theory is understood, and as a preparation for it. In that case, however, glass would not be at the beginning of a series of object lessons; it would need to be subse-

quent to 'sand,' 'ashes,' 'soda,' and also 'heat,' in one of
its more recondite applications. This is an example of
the troubles of the object lesson, at the beginning ; the
thing chosen may be familiar, but what is of interest to
add to it may bring in something very abstruse. There
is but a choice of difficulties. Confine yourself to what
the pupils know, and you teach nothing; endeavour to
extend their knowledge, and you land them in the unintelligible. Every street-Arab knows all about 'glass,'
and has a great deal of other knowledge, which perhaps
occupied many of the school hours of the well-trained
youth.

The only mode of escaping this alternation of difficulties, is to look before you leap—to see beforehand
which way you are going, and whether or not your way
has been already prepared. At the absolute commencement, you are stopped on every side ; still, it is expedient to make some move, and the safety lies in moving
only a short way, in drawing but little upon previous
knowledge. This very proper caution, however, does
not fully meet the case. The real remedy lies in *pre-
arranging a set of lessons*, such that each shall be a preparation for the following, and in guiding the course of
the tuition by reference to what has been already taught.
This cannot be done with perfect rigidity, at the age of
desultory knowledge, but it can be done in some degree.
A substance might be introduced at one stage, and
followed out just as far as previous knowledge permitted ; it might be re-introduced at a later stage, with
new expansions. 'Glass' at first would be noticed
merely for observed uses and properties ; to these very
little would be added. At a subsequent stage, its

manufacture could be propounded; and still later, its optics.

The second essential of the Object Lesson is a definite purpose, a limitation of scope. The teacher should consider what is to be the drift of the lesson. That, at the outset, lessons are more or less desultory, perhaps cannot be helped ; but they should gradually be brought under some of the ' Unities.' Now the purposes are various, and should be distinctly grasped. A reference to any of the usual examples of Object Lessons will show the danger of putting too much into a single lesson ; while, apart from a very strict consideration of the unities, merely keeping the new information within limits of quantity would render it safe.

Let us take one of the usual examples—a bell. For very young children, this may be little more than an exercise in observation and description. The prelude is the incident of being called to school by the bell. Next, a bell is shown ; probably most of the class have had one in their hands. They see the cup shape, they notice the clapper hung inside, they see it swing, and knock the cup, and with that comes the sound. It would be quite enough for one lesson in the early stage, to trace cause and effect in sound by the knocking of one hard body on another ; adding a few parallel facts gathered from the pupils' own experience, and brought out by questioning. This of course is nothing that they would not ultimately know of themselves; but, by being brought in early, it may be a stepping-stone to more recondite truth ; in fact, a first step in the ladder mounting to Acoustics. As to the many occasions when bells are used, that belongs to the popular and amusing essay, and does not lie in any line of mental discipline. Even the metallic structure is premature, although, at a later stage, it may come in as explanatory of the loudness of the bell. The lesson is managed simply as a lesson of cause and effect, in the empirical form, and although such a lesson deserves the name of Science, it does not pass beyond the interest and comprehension of the child of seven years.

A piece of chalk, as already remarked, has been considered as a worthy Object theme for an audience of full-grown people. Many sciences centre in it, and therefore it can be the starting-point of an agreeable excursion in any one of several lines. It is implicated with Zoology, Geology, Chemistry, and Physics, and may be made the occasion of stating or recalling interesting truths, in every one of these subjects; all which truths are lodged in the memory by their connection with it. It is also implicated in numerous utilities and processes in the arts. There could not be a better example for the teacher, to be put forward by him on successive occasions; a limited purpose being kept in view in each. The Zoology and Geology should obviously be very late; either after these subjects have been partially introduced, or with a view of introducing them for consecutive handling. What could be given separately as an early lesson (which the Arab would only by rare chance attain to), would be the burning of chalk and its equivalent, limestone, in a kiln, yielding quicklime, to be afterwards converted by water into slaked lime, and then used with sand for mortar. A strict statement of these circumstances, without any digressions, would be an interesting chain of empirical cause and effect, to be one day used in expounding chemical and physical forces.

When a substance is quoted solely and simply for its *use*, other things having the same use may be quoted; the lesson is then a *generalizing* lesson, and the remaining circumstances should be put on one side. Thus, if coal is introduced to teach combustion and heat, other combustible substances may be mentioned—as wood, rags, dried leaves, sulphur. No other facts about coal should be adduced in this connection, except perhaps in the comparison with wood, when the common origin might be just mentioned. The topic of the lesson consisting in the *single fact of combustion*, all further reference to the properties of heat should be foreborne, as belonging to a distinct lesson.

Again, a lesson exhaustive of the uses and properties of a substance, should not pursue any one property either by expounding its laws, or by quoting all the other things possessing the property. The end is, to give a full account of all the characters that concur in one substance—to group or totalize its powers and uses. This admits of nothing beyond the bare mention of the various uses, with only enough explanation to make them intelligible. Thus, Lead is one of the metals (two or three others being merely mentioned), heavy (ten times water), soft (for a metal), ductile, melts in an ordinary fire, does' not rust like iron. It is used for making pipes and cisterns, for bullets, for solder. The uses might to some extent be connected with the properties, but to do so is to trench on lessons of property as cause and effect. This is an incipient lesson in Mineralogy, and should be followed up by some other metals, treated in the same style, and by a few substances not metals. There might intervene lessons following out distinct properties throughout different substances, as weight, ductility, corrosibility; the contrasting or negative examples being also quoted. By such lessons the properties would be more fully comprehended, when spoken of in connection with any one body; but the two kinds of lessons should never run into one.

Lessons on flowers, plants, shrubs, trees, as exemplifying the vegetable kingdom, come under the principles now illustrated; they should fall into an order tending to some definite purpose, and each lesson should be a unity. When first adduced, specimens of flowers and plants serve a sufficient purpose, if they bring out observation and verbal description. For, although they may be familiar in general aspect, the pupils have not observed any one thoroughly. This they may be taught to do, and also to name each part of a plant—root, stalk, branches, leaves, flower, seed, and so on. Different plants may be used, for the discrimination of the parts simply. Then comes a lesson on a particular plant—a daisy, for example—to mark the forms that root, stem, &c., assume in it. Several others, including known shrubs or trees, are next adduced. Then might come a *generalized* lesson on trees, grounded on a few known examples; going no farther in the first instance than to indicate the notable features of magnitude, strength, hold in the ground, branching, leaves, and flowers. The growth, maturity and death of trees, would need several separate lessons; and distinct from these would be the sources of nourishment, by the roots and the leaves;

which would be an empirical lesson in advance of the science of Vegetable Physiology.

Before including animals among the examples, I will give the third law of the Object Lesson, which has reference to its use in adding to the store of concrete conceptions : this is commonly expressed by saying that it cultivates or enlarges the Conceptive Faculty or the Imagination. Basing upon what the child already knows and conceives, unknown objects may be pictured forth, and so laid hold of, as permanent imagery for after uses. It is thus that children may be made to conceive in a dim form, the camel of the desert, the palm tree, the Pyramids of Egypt. Now, as far as the power of conceiving goes, there is not so much evil in being desultory; a rambling style may favour the culture of this faculty. Anything that makes an impression, makes a recollection.

But a teacher may readily overrate his power of adding to the stores of concrete conceptions by means of description, and may still more readily mistake the bearing of the object lesson on this acquirement. The earliest display of the power of obtaining new conceptions of things not experienced is strongly, not to say grossly, anthropomorphic ; and is the result of piquant narrative. A cold lesson on lead, or on glass, on a soap-bubble, or on clouds, does little for the power of conceiving the absent concrete ; its chief agency is to impress the known more fully and clearly, and thus to prepare for the future operation of figuring the unknown. Indeed it is only at a very advanced stage, when the object lesson is swallowed up in the methodical study of Geography and History, that it can be properly mentioned as an aid to the increase of conceptions formed by the mind's own force of combining the unknown out of the known. The only seeming exception to this view will come out presently, in dealing with the examples taken from animals. It is in this class that the licence of digression runs wildest. In proportion as the characters of the humblest animal exceed the utmost that can be said of either mineral or plant, the teacher's selection needs the control of a methodical procedure.

The first introduction of animals in lessons generally turns upon their broad mental characteristics, which are intelligible to every child ; their search for food, their victimizing other animals for the purpose, their amiable traits as regards their young, their human attachments. A short narrative framed to bring out these, with the interest of a plot, is both engrossing and impressive, and is readily received by the memory. By virtue of such interest the form and physiognomy of animals is stamped on the mind. If the teacher is cautious, he may make a start from here, and travel into some of the minutiæ of the natural history of the animal, as its claws, its teeth, its hair, its wool, its feathers, and may render these still more familiar. To proceed beyond this point, he will have to make a choice of ways, exactly as with the plant, but with still greater *embarras*. There is the same amount of peril, in both cases, the attempt at comparison with other animals, either generically related—as the cat, tiger, lion—or more distant— as when the cat and dog are compared. Comparison should not begin without adequate preparation, that is, without the previous mention of the most suitable instances ; and when it is made, it should be rigid, thorough, and to the point: it should aim at forming a class, ·with class attributes, to the neglect of the differences between the several members of the class.

As in the plant, the other method and the method that is prior in order, is Individuality, or the mention of particular characters, without running comparisons or contrasts, and with only so much expansion as amounts to being intelligible. In describing the rooks and their rookery, for example, it is very well to state their manner of feeding, their coupling to build their nests, their associating in multitudes and behaving like a society. All this belongs to the individual subject ; but it is a misplaced digression to be led off to social animals generally, as bees, ants, and beavers. That is a theme by itself—the theme of Generality—to be taken up after due preparation. It should be preceded by the detail of the most remarkable examples, and discussed solely with a view of comparison and contrast of the different species.

In the individual descriptions, a passing allusion may be made to another species (especially if already brought under the notice of the pupils), but it should be merely illustrative of a meaning, and should not be pursued.

To exemplify the several maxims governing the Object Lesson, I will take the example of the Camel. This animal has not been seen by the pupils, but they will be shown the picture of it. It should not be an early example. The more familiar home animals of the useful and domesticated kind,—including, the horse, ass, cow, sheep, deer, &c.—should precede. We are not bound to the strict order of a Zoological description, yet there is a method to be observed in bringing forward the points. First, the camel may be designated as a beast of burden ; that is not only a comprehensive circumstance, the key to much that follows, but it recognizes use as property in act. A very briefly stated comparison may be allowed to other animals of the same use—horse, ass, reindeer, elephant ; but the property is not to be dwelt upon as if it were the theme of the lesson. The interest of the animal turns upon its structural adaptations to a peculiar situation, namely, the desert. Here we have a *double subject*, with mutual bearings ; it is a case of correlation where the order is not absolute. We may, however, begin with the situation, that is, with the desert, but *may describe it only so far as it concerns the camel;* we may give the facts or features, without the whole chain of causation, which is a quite distinct lesson, belonging to the strict sphere of Geography. 'In many parts of Africa, Arabia, and Syria, there are large tracts devoid of water, and of vegetation, except at long intervals, the surface being dry sand or naked rock ; the occurrence of water accompanied with vegetation makes what is called the "oasis" in the desert.' It is quite extraneous to mount to the causes of the water, in the deficiency of rainfall, owing to distance from great oceans, and so forth. Next is the form and structure of the camel. The singularity of the hump concentrates a part of the external description ; so its growing smaller in the absence of food, as being a reserve of sustenance. Then comes the stomach, which in general make resembles the stomach of the ox, the sheep, the deer (called the ruminating stomach), but differs from these in being able to store food and water for long periods. Feet spreading, and not compact like the horse's ; thus fitted for the sand. The eye protected from the sand that blows up in the desert. The knee adapted to kneeling down for the reception of its load. All this description has its interest and relevancy solely from the point of view of use. The naturalist's description would be far more exhaustive, and would comprise points that have no ob‑vious adaptations.

Hitherto we have seen in the Object Lesson a mode of approaching the Natural History Sciences, as Mine‑ralogy and Botany; we shall afterwards see its application to Geography and to History. The three maxims that have been exemplified, namely, (1) Sequence, (2) Indi‑viduality, and (3) Generality, are directly pointed to this class of lessons. But the Natural History Sciences lead up to the Primary or Fundamental Sciences—Mathe‑matics, Physics, Chemistry, &c., in which are found the final explanation of all the active agencies of nature ; everything expressed by power, force, causation—the laws of Motion, the forces of Gravity, Heat, Electricity, Vitality, and so forth. We do not know the phenomena of nature, until we know them as produced and pro‑ducing according to their general laws.

Natural History descriptions contain a tacit reference to these higher powers. A mineral has specific gravity; that implicates the great power of Gravitation. It has transparency and refracting power ; that implicates Heat. It has composition, which implicates Chemistry. But the mineralogist knows his business ; he merely alludes to these great powers, he does not set them forth in me‑thodical exposition. The reserve in this respect is not always copied by the Object Lesson naturalist ; there is a tendency to rush on from the natural properties as de‑scriptive characters, to the full exposition of their work.

ing,—to make Natural Science absorb the Primary Sciences.

It is possible, by means of the Object Lesson, to approach the Primary Sciences, namely, Physics, Chemistry, and the rest, with the view of explaining Matter and Motion, Gravity, Heat and Light; but the manner of doing so needs the gravest consideration. There must be a clear disentanglement from the Lessons on the Natural History type, whether Individual or General ; all the more so that there may and must be points of contact with these, because the same concrete things enter into both.   Thus, for instance, Lead may be the basis of a lesson in mineralogy, whether individual, as exhausting its properties, or general, as under the class ' metals '; but it also comes forward in physical and chemical Science, under Gravity, Heat, Chemical Combination, and so on. In this last case, however, it is merely one example of a countless number of things that are equally suitable for expounding the great physical forces ; Gravity, Heat, and Chemistry have an unlimited choice of examples to show their operation.

It being assumed that the best and only perfect way of explaining the primary sciences is on their own methodical plan, as laid out in a course of Physics or Chemistry, the question before us is how to manage those interesting anticipations of the leading doctrines, fitted to the age when the regular course cannot be understood, paving the way for that course, and making up a body of information valuable so far as it goes, even if the pupil never passes through the final curriculum.

So great and manifold are the advantages of following the regular order, that the teacher should always be looking forward to the time when the advancing intelligence makes that possible. And further, he should tacitly keep this order in his mind, even when working on the seemingly desultory plan. Thus, among the earliest lessons that implicate physical doctrines, Motion, as exemplified in visible bodies, should have a chief place. The force of Gravity should precede the more subtle forces of Heat and Magnetism.

When we enquire farther into the principles regulating this kind of teaching, we find that the lesson belongs to the *empirical* form of knowledge; the meaning of which is, that facts are stated fully, faithfully, correctly, but not explained or referred to the ultimate principles or laws that they come under. The phenomena of the Tides can be described fully and correctly in the empirical form, as it was known before Newton ; while in the early lessons in science, the pupil cannot be made to under stand how they arise from gravitation. The statement may be given that they arise from the gravitation of the sun and moon, but this cannot be fully shown, or correctly conceived, except by the pupil that is pursuing Astronomy in regular course, after a due mathematical preparation. It is merely confusing the mind to assign a cause in vague terms, as Gravity or Electricity, when it is not possible to make the working of the cause intelligible. Little good is done by saying that thunder and lightning is a fact of electricity, when electricity itself is not understood. Still, an object lesson in thunder discharges might be given, which would comprise the main visible circumstances, together with the atmospheric accompaniments and surroundings, in so far as conceivable by the pupils addressed. The antecedent

circumstance of excessive heat in the weather, the gathering of the dark cloud, the deepening of the gloom,
the lightning flash, or thunderbolt, often the destruction
of buildings and animal life, the booming of the thunder,
at a varying interval indicating distance, the deluge of
rain—might all be described, partly recalling the experience of the pupils, partly awakening their minds to
watch the next storm, partly extending their own observations by depicting the usual forms of the lightning,
and stating instances of its effects, but not embarking
upon the theory of atmospheric electricity, nor even
naming it, further than to say that they will at some
future time be made to understand a great deal more
about the phenomena. Whether or not the teacher
should use the opportunity of bringing forward the somewhat easy and yet interesting and intelligible fact that
sound occupies time in reaching our ears, depends
upon the course of the tuition. Such a fact might be
previously brought forward in a lesson on Sound and
Echoes; if this were so, it would receive a passing
reference, and an impressive exemplification in connection with the lesson on Thunder. But as regards any
lesson in Primary Science, the great caution is against
overloading; the pupil must not be led to suppose that
there is but one chance of explaining half a dozen natural laws stretching out into several sciences. Because
all the sciences meet in Water, that is not a reason for
embracing them all in a single lesson, nor indeed for
attaching them to that one object. The laws applicable
to water are applicable to a thousand other substances;
many of these sufficiently familiar. The Tides might be
given as a water lesson; but we may just as easily start it

under the name 'Tides,' as under the name Water. The most suitable designation would probably be 'The Tides of the Ocean.' Its regular place would be somewhere in Physical Geography ; but it might be given at a still earlier point in the child's course.

When choosing an Object Lesson, we should think more of the principles to be taught than of the text ; the selection of the text is only the second consideration. We must not be dominated by our text Object. We may make the Ocean the text for a lesson on the Tides, but we are not to be led off into facts regarding the ocean that are unconnected with the special phenomenon of tidal action. There is a unity in the subject of the 'Tides' ; the unity belonging to an Object Lesson in the primary sciences—a phenomenal unity. There is no unity in the subject of the Ocean, until we have first determined what use we are to make of it.

The texts suitable for the present kind of lessons are given by a class of names different from the names of the two foregoing classes. There are names that point to natural objects—as water, iron, an oak, a horse, a star, a mountain ; such are the starting points of the previous lessons—those in Natural History, Geography, and the like. There are other names that call to mind the processes, powers, and operations of the world—as weight, heat, dew, attraction, polarity, respiration ; these are the names that give the most convenient start to the science lessons that we are now considering: although any one lesson *might* be associated with the more concrete names. A heat lesson might begin from water ; an electricity lesson from iron : but this is not the course to be recommended ; it has a false glare of simplicity. Each lesson should be taken for what it is, and connected with the name that best indicates and circumscribes it.

The 'Atmosphere' is a common example of the Object Lesson. It cannot be called a happy or a convenient starting point. Nothing could be a worse policy than to attempt to exhaust (if it were possible) the natural facts implied in it—its physical, chemical, and biological relations. We could merely nibble at them ; we should teach nothing thoroughly ; not to speak of the evil of perplexing the mind of the pupil The proper use to make of the Atmosphere, as a text would be the Natural History use; it would be an Individual or concrete lesson, whose properties or peculiarities should be simply enumerated as Natural History. Beginning with its position on the earth's surface, we might give its supposed height, its mass or weight, its gaseous character, its transparency. We then go on to its composition, which would require us to enumerate Nitrogen, Oxygen, Water, &c., with perhaps a word or two to render these as intelligible as the state of advancement of the pupils would allow. We should certainly reserve all questions connected with the origin of the water constituent or vapour, which would carry us out of the lesson into a totally different track ; we could simply mention briefly that the water constituent was of variable amount, and, while in. great part invisible like the others, had visible manifestations in clouds and mist, ending in rain. No more of that, if we mean to finish the lesson in its Natural History type. We then go on to a similarly guarded and severely curbed enunciation of the carbonic acid constituent—its amount, its character (as the gas formed by the burning of charcoal, wood or coal), its function in supplying food to vegetation. There would still remain the smaller constituents, including the effluvia of the earth's surface, animal germs, &c., which could be simply mentioned, without being pursued.

To penetrate deeper into the mysteries of the atmosphere, to trace the numerous laws of causation involved in it, the lessons must follow other tracks, and be viewed in wider connections. An example or two will help to explain our meaning. The primary property of the atmosphere is the fact, not apparent at first glance, that the air is material and inert like the visible and tangible bodies around us. A very good object lesson might be contrived to exhibit this circumstance, which possesses the interest of agreeable surprise. The proofs and illustrations from resistance of the air, wind, and so on, are well known and highly impressive. But this lesson would really be a lesson on the inertness of matter ; and would in fact have for its proper designation—Matter and Motion. As resistance to our energies, exhibited by solid and liquid masses, would be the first circumstance of the lesson, the illustration would be naturally carried out to air, thus establishing the material quality of the air. Then as to the weight and pressure of the atmosphere, there would be a natural alliance with a lesson on Gravity or Weight, which might be made intelligible at an early stage, although still in a considerable degree empirical. It would not be among the earliest lessons of a scientific tendency ; for its adequate handling would presuppose the globular form of the earth, and some general conception of the solar system. Next to the weight of air, is its elasticity;

this would come under a mechanical lesson on Elastic bodies or springs ; from such a lesson we ought.not to omit the spring of the air. Yet we could not properly follow out, in the same lesson, the interesting consequences of the spring or elasticity of air combined with gravity, as the rarefaction of the air in the upper regions; this would want a lesson to itself.

The constitution of the atmosphere as made up of Nitrogen and Oxygen appeals to Chemistry, and to Chemistry we must go, but on some other occasion. For the present lesson, Oxygen receives a few suggestive touches, yet only in empirical statements, shaped according to what is known of the pupil's previous course. At best such statements are incomplete and unsatisfactory, if not even misleading ; the only safeguard is, not to be carried away by an attempt to explain them.

The water constituent of the atmosphere, with its wonderful transformations and its perpetual cycle, is, if we may judge from the lesson books, a favourite topic of object teaching. The one fact of Dew is the more especial favourite ; although in point of difficulty, it makes a very advanced lesson in Physics, as taught in a regular course. This is a good case for exemplifying what to do and what to avoid in the Object Lesson, and may help us to see the necessarily empiric character of the early scientific teaching.

Because the teacher is debarred by the capacity and knowledge of the pupils from a scientific lecture, it does not follow that he should be incapable of giving such a lecture, or be ignorant of the place that the subject would occupy in a connected syllabus of the Sciences. It is far better that he should know this, in order to know why and how he is to depart from it. Thus, in a course of Natural Philosophy or Physics, 'Dew' is explained, under Heat, which subject is preceded in the course by Dynamics, Hydrostatics, and Pneumatics. A wide basis of physical knowledge has thus been laid in the mind of the regular student; in particular the laws of motion, and the law of gravity, have been applied fully to solids, liquids, and gases; while, in the subject of Heat, where Dew comes in, some of the leading facts have been expounded, as the expansion of bodies, lique-faction, and vaporization, and their opposites, with the doctrine of latent heat. Stored with all these prepa-ratory explanations, the regular student of Physics is introduced to the topic of Dew; and the teacher still finds a good deal to say before it is completely mastered by a youth of average intelligence. Taking all this into account, we should naturally despair of bringing before pupils of ten a subject that fairly tasks the powers and the acquired knowledge of a pupil of sixteen. Such would be our first thoughts. The second and better thoughts are to consider what limitations, omissions, precautions, the altered circumstances impose upon such a lesson. We begin by stating to ourselves the reasons for making the attempt at all; namely, to engage the attention of the young mind with the facts, appearances, and operations of the world, so as to have some impres-sions that the regular teacher can afterwards work upon; for the professor of Physics, in his lecture on Dew, would be very much at a disadvantage with pupils that had never even noticed the wetness of the grass on a morning after a clear and rainless night. We next recall the cir-cumstance, that cause and effect, in some form or other is noticeable by and intelligible to the youngest capacity, and even seizes hold of the attention of its own accord; nay, more, that the youngest mind will form an induction to itself of the conditions of any startling change. Every child is a self-taught natural philosopher in such matters as the fall of rain, the wetting of the ground and the filling of the water channels; and will reason, from the occurrence of wetness and rushing streams, that rain has just fallen. To guide, rectify, direct and forward this spontaneous observation and reasoning is the purpose of the teacher in the lessons that we are now considering; with the serious drawback, however, that the perfect form of the truths cannot yet be imparted; and that, on the way to the perfect form, the pupil has to pass through several forms that are imperfect.

Before applying these reflections to the case of Dew, the remark is significant and helpful, that a century ago Dew was not understood at all; until Black had expounded latent heat, and Dalton studied the constitution of the steam atmosphere, no satisfactory account could be given of the phenomenon. Still it was not entirely unknown, and such knowledge as was possessed was correct and useful. This shows us that there are forms of knowledge, short of the highest, that yet possess value. That former knowledge of Dew was *empirical* knowledge; and the knowledge that we give to children in advance of the perfect form we have designated empirical too. It is so, however, not by the necessity of the case, as it was to our fathers, but by deliberate and artificial shaping on our part. We know the real solution, the rational explanation; but we withhold it as premature.

Yet, here is the advantage of our position; we can use our full knowledge to improve the empirical statement, to make it less removed from fact, and more full and intelligible for its immediate purpose. We can let drop forecasting hints as to what the pupil will one day fully understand; we can even tell the real cause in a general way, while we cannot point out all the steps. It does no harm to complete the empirical account of the Tides by the indication that they are due to the united attraction of the Sun and the Moon; our error is to attempt to show this in the detail to pupils that are incapable of abstract dynamical conceptions. We can give them a very valuable lesson without an over-vaulting and premature attack upon the citadel. We engage attention and observation upon a great terrestrial fact, we plant a large conception in the mind, we give a *proximate* explanation of a phenomenon of perpetual occurrence; we sum up in a generality a host of scattered appearances: we are thereby justified in putting forward the subject as a knowledge lesson in advance of the pupil's attendance in the Natural Philosophy class room.

To resume the example of Dew. While the lesson is avowedly given to those that cannot understand the reasons or explanations, and, therefore, does not presuppose all the knowledge that should properly go before, it still needs some previous preparation of mind, and must take shape according to the supposed knowledge of the class. It ought not to be given without certain other lessons; such as, the materiality of the atmosphere; the three states of matter as depending on heat—a very good example of an empirical lesson; the boiling of water; the difference between gaseous water proper and visible vapour or steam; the drying up of wet surfaces, and of ponds of water; the heating of the air by the heat of the day, and its cooling at night. Such points being premised, the lesson might assume this form: – Water, when disappearing by drying, becomes a gas diffused in the atmosphere. The atmosphere does not hold above a certain quantity. What is the consequence? Either the drying must stop, or it must be thrown down again to the earth. It *is* thrown down in the form of water as rain. This is the chief mode of returning to the earth. Before it appears as rain, it exists as clouds, which feed the rain. Rain comes when the air is cooled by the vicissitudes of day and night, and by changes in the wind; the great fact is coldness. We can obtain water from air in various ways, if we cool it enough. The ground becomes cold at night, and the surface is wetted, although there has been no rain.

The sum and substance of the lesson would be to connect drying with the heat of the air, and the return to water with its cooling; to impress which in broad general terms would be quite as much as could be done in one lesson. Obviously, the rain and cloud lesson should precede the lesson on Dew, which is an exceedingly subtle consequence of the general fact. The reasons why dew is absent altogether on some nights, and why in one night some bodies are dewed and others not, cannot be imparted intelligibly without a distinct lesson. The statements might be given as *empirical facts*, that grass and wool are more liable to be dewed than stone and metal; but the theory of surface radiation and of its differences in different bodies should not be foisted in for the first time in a Dew lesson ; either it should have occurred in a previous lesson, or it ought to be entirely withheld, leaving only the empirical statement. It is the very essence of the Object Lesson to be empirical.

In an Appendix note, the niceties of the lesson in Primary Science are further brought out by a critical review of some select examples. To this is added a discussion of the forms assumed by the Lesson as arising in the explanation of words that occur in the reading books.

Geography.

The aims of Geography are very well-defined. The conception of occupied space is its foundation ; it is the all-embracing framework of the outer world in its orderly arrangement. On the great scale, it gives a place to everything, and peoples every place. It is the greatest task of the pure conceptive power, in its literal or matter-of-fact working, as opposed to the imaginative or emotion prompted working ; this alone would make it a late study, as the child has but little concrete conceptive faculty, and that little is disturbed by the intrusion of strong emotional effects.

A long series of lessons on the isolated objects of the outer world—implements of utility, minerals, plants, and animals—serve as part preparation for the vast geographical field ; but that field opens up an entirely new exercise of the conceiving power, which must be grounded on a distinct line of observation and experience. The simplest objects of Geography—hills, rivers, plains, occans, cities—are immense aggregates, while the idea of the science is, to seize in orderly array the multitudes of these that make up the surface of the peopled earth.

For introducing the elements of Geography by means of object lessons, the chance impressions of a child of eight or nine seem wholly inadequate. It would be necessary to take the class out of doors, in Saturday excursions, to mark with express attention the surrounding scenery in its comprehensive aspects, and to conceive the town or village, as a whole, with form and parts. It is from some commanding eminence that a pupil should receive first impressions of Geography, if the subject is to be taught according to the prevailing wish for concrete realization. In a district that is flat and monotonous, like our Eastern counties, there is scarcely the material for geographical conceptions; while to vast numbers of people, a notion of the sea, simple as that notion is, is utterly debarred. Few are unpossessed of some notion of a flowing stream, by which to conceive a river as a moving body of water; but the geography of a river in all its expansion, demands previous acquaintance with mountains, valleys, plains, and seas.

Inadequate and difficult as the propaideutic may be, it is creditable to the schoolmaster to make the attempt to force attention upon the actual surroundings of the pupils, and to work these up into conceptions of other places differently arranged,—to use the experience of sunshine and rain, of heat and cold, of snow and ice, for conceiving countries where the hottest days at home are the constant fact, and others where ice and snow endure three parts of the year. All this is the legitimate culture of the conceptive faculty, as a means of knowledge and truth ; the chief error to be avoided being the premature entry upon a very high accomplishment.

In adopting, for a lesson, any one of the great geographical elements—for example, a river—the laws or method of the object lesson need to be very narrowly observed. The greater difficulty and vastness of the conception requires still more peremptory attention both to sequence, and to the unities. The point of sequence has just been touched upon, and ought to be more self-evident for this kind of lesson than for those already described. The adherence to unity of plan, as against temptations to digress, will ever be the hardest task of the teacher in object lessons, and it is most of all requisite in Geography. Thus, in the example of the River, one distinct lesson, and indeed the main lesson, in the geography scheme, is to conceive the visible aspect of the flowing waters, in the main stream, and in all its branches, from the first rills emerging out of the oozy hill tops and hill sides. To make up one visible picture of a river tree, as if from a bird's-eye view of its entire basin, all collateral explanations must be resolutely withstood ; and if the first source of the whole—the rain —is mentioned, it should be no more than mentioned, while all the numerous relations of rivers to the fertilizing of the land, the supply of water to cities, navigation, and so forth, should be omitted from the primary lesson. Hill and valley are already assumed, and the river located with reference to them ; the final debouching in the ocean is to be mentioned without being followed into any of its consequences. It is enough for a week's lesson, by iteration and examination to stamp the mere visible plan of a typical river with its tributaries, brooks, rivulets, and

cascades. All comparisons and contrasts should be re-
manded to a lesson, or lessons, on Rivers *as a class*, with
class agreements and differences. The other excluded
topics, and forbidden digressions, are matters pertinent
and proper to be known in connection with a river; but
each has a place and connection suitable to itself. The
ultimate source of rivers—the rainfall—belongs to the
department of Physical Geography, or else to the physical
science of Meteorology. The use of rivers in draining
off superfluous water at some points, and supplying
water at others, is quite a different department, and may
be subdivided into several topics. The connection of
rivers with towns, as ministering to numerous wants and
conveniences, comes in for full treatment at a late part
of the subject, although passing allusions may occur
in a variety of the early lessons, as under 'water,' which
is a point of departure for numerous lessons on the object plan.

Cause and effect is at all times an impressive circumstance; but we have seen that the efficacy of causation
lies in bringing about an effort of abstraction, which
interferes with the concreteness of the visible picture.
It would be well, once for all, to attain a good pictorial
impression of a river basin, as it is spread out to the
actual view, before, and apart from, contemplating the
numerous exemplifications of causal agency that it supplies; all which, when known at another stage, may re-
act on the concrete conception, by supporting some of
its constituent notions, without dissolving the picture.
Thus, assuming the rainfall as the ultimate river supply,
the influence of rainy weather in swelling all the afflu-

ents, and in enlarging the mass and impetus of the final stream, is highly operative as an aid to the picture.

A Town is a very suitable object lesson, at an early stage, as contributing to Geography and to other purposes. There should be the same adherence to the visible or pictorial conception, in the first instance ; the same avoidance of digressions, until that conception should be firmly fixed. Subsequent lessons, returning to the subject, could deal with the reasons of town arrangements, and with interesting details, under specific heads ; while the comparison of several towns would exemplify the standing lesson of expounding a *class*, by agreements and differences.

The pictorial view of Geographical subjects is supposed to be aided by sketches from nature. But here we encounter a new danger- the supplanting of the original reality, so arduous to the feeble conceptive powers of childhood, by the more easily comprehended sketch. Young and old alike, on seeing a good picture, are apt to rest there. Such is the tendency of all representations, sketches, maps, and the rest. The most valuable helps to Geography are models, and if these could be multiplied in schools, the conceptions of the general form of countries would be vastly enhanced ; while the subsequent lessons of juxtaposition and relative situation would find a groundwork of remarkable cohesiveness.

In the manuals of Teaching, much is made of the necessity of introducing the pupils to the meaning of a Map, by showing plans of the school, and of surrounding things known to them. They are in fact too ready to accept the map as the subject-matter of the future lessons ; on it they can see the situation of countries, the course of rivers, the outline of coasts, and everything that they are expected to give information upon; and to make them rise from the map to the actual conception of the ground is an attempt too difficult to sustain; it can be done but rarely, and under special helps. What the map cannot show, the pupils learn in verbal statements, which are remembered as such.

The Compass is an easy object lesson, implicated as it must be with the course of the sun. It is a higher stretch, but by no means difficult to children of eight or nine, to grasp the foundations of latitude and longitude in the form of the earth, which may be quoted for this purpose, without involving anything beyond. The idea of laying out a surface into compartments, by cross lines at equal intervals, is sufficiently easy, and is important as a key to the orderly arrangement of the things contained.

At all stages of Geography, local situation, form, magnitude should be a distinct effort of memory, and should be held in the mind as visible facts, grounded on the map or model. The impression may be very much assisted and strengthened by the varied information relating to cause and effect, and the mutual relationships, but these should not be given until a certain hold of the visible order on the map has been secured. There is one important rule in all teaching, viz. to separate the fact from the reason, and to describe the fact first, that it may be understood and imbibed as such. This applies to the vast subject of geographical relationships; the surface is given first as a fact, and afterwards considered in the numerous links of dependence existing among the different elements that make up a tract of country.

The art of laying up in the memory geographical positions involves very delicate manipulation on the part of the teacher. The method of proceeding is that embodied in the rhetorical principles governing Description ; the chief being to start with a comprehensive plan or outline, and to subdivide the whole into parts, either at once, or by successive divisions, as the case may require. At the stage of progress when pupils are expected to take up the map of Britain, they are competent to view the globe as a whole, with its divisions into continents and seas, and to descend by regular subdivision to our own country. The operation is as easy on the large scale as on the small.

The School-books give in unexceptionable order the topics to be stated in connection with the map of any part of the earth—larger or smaller. Of only recent introduction is the method of describing, that pictures out the surface in orderly array, dividing it into mountain ranges, valleys, plains, and giving these in their proper positions. The system was exemplified on the great scale by Ritter, and first carried out in this country in the 'Penny Cyclopædia.' It has since found its way into the smaller manuals, the earliest to adopt it being the manual of William Hughes. When the pupil is sufficiently advanced for a manual of this kind, the teacher's path is made quite plain. The following out of the position, boundaries, form, magnitude, and general aspect and features of the country, into the consequences entailed by these on the vegetable and animal products ; the enumeration of those products ; the account of the inhabitants, and their industries and social condition (Political Geography),—are all sufficiently well done in many text-books.

The science called 'Physical Geography' is something intermediate between ordinary Geography and the higher sciences, namely, Physics, Chemistry, Meteorology, Botany, Zoology, and Geology. It introduces considerations of cause and effect into Geographical facts, by selecting and stating in empirical form the principles methodically taught in the regular and fundamental sciences. A course of Physical Geography is subsequent and supplementary to proper Geography, while reacting upon it in the way that causation operates upon the knowledge of facts. It is also an introduction to the mother sciences; but until the principles are studied in their due order and dependence in these sciences also, they do not leave their mark behind them.

The teacher is tempted, now and then, to bring under the proper or descriptive Geography the scientific explanations of physical geography. Any such explanations should be very short and allusory ; the two departments should be by no means intermingled.

There is a still greater temptation to include History with the descriptive Geography. This serves a purpose in rendering more intelligible and interesting many of the facts, especially of Political Geography. It should, however, be very shortly and sparingly done; being confined to the exact purpose of aiding Geography proper. Attention is properly called to features that determined great historical events, as a preparation for the study of the history, but without dragging in the history there and then. There is a separate branch of knowledge, falling under Political Philosophy, or Socio-

logy, which traces the dependence of the social arrangements and social development of mankind upon physical
circumstances.   An interesting and salient fact taken
from this department may be occasionally noticed in
geographical teaching, but the department as a whole
cannot be absorbed into School Geography.   Like Physical Geography, it must have a place in the curriculum
all to itself.   At the same time, it is a merit in the Geography teacher to forecast this application, and unobtru-
sively provide for it.

In Geography, much has to be learnt as words, or
little more ; the verbal memory has a large share in
the acquisition.   In this view, the names should be relieved of dryness by various arts, as well as by endeavouring to impress real conceptions corresponding to
them.   Yet we must not overrate the conceptive power
of young pupils, in a subject that in a great measure
excludes the strong emotions.   That a youth of ten
should conceive the plains of India, with their vertical
sun, their peculiar vegetation, their animals, and their
swarming dusky population, is not to be supposed.   The
best arranged series of object lessons cannot prepare the
mind for all the characteristic plants and animals of a
tropical region ; while the constructive effort that gives
them their places in the landscape is possible only in
the full maturity of the mind, and is even then attained
by a very small number of persons.

Geography may in various ways be connected with
the exercise of drawing.   The drawing of maps impresses a country, just as copying a passage in a book
impresses the author's language and meaning.   In those
cases where drawing is followed out as a fascination it

carries with it an interest in the face of nature, and an enhanced power of conceiving the pictorial aspects of the world. In addition to which, the influence of poetry may come in aid of the geographical concrete. Tennyson's 'Brook' is the rendering of one of the numerous affluents of a mightv river.

HISTORY.

The transition from Geography to History is natural, when History is conceived in its highest or final form. But, as a subject of teaching, History passes through many different shapes. In those early narratives that seem indispensable to the interest of the first reading lessons, being almost the only device for riveting the attention of the very young, we have the initiation into history ; indeed, the persistent catering for stories brings the teacher at last to actual History, through the intermediate stage of Biography. In the lives of kings, statesmen, generals, and other great men, we have the materials of history.

The full bearings of History cannot be understood without much previous knowledge, and some experience of the world ; and where these requisites are found, there is little need for a teacher. The great historical works, ancient and modern, are the self-chosen private reading of our mature years.

The earliest lessons of a general kind, in connection with History, are lessons in human nature, in the ways, actions, and motives of men. These may be very elementary and obvious, as in the displays of selfishness, of devotedness, and of the various other forms of human passion. When such passions animate a nation, or a collective assembly, they are facts of history. It is desirable, however, to bring before the older pupils the exact nature of *society*, as an assemblage of human beings in a certain fixed territory, for mutual interest and security, and presided over by a head, or governing power. Out of this arrangement comes law or social obedience, which is a great part of morality, and the type of the whole. History presents different forms of government, and different kinds of laws, and in its narrative portion contains the changes more or less violent in the relations of the governing body to the governed. When such matters as these are exemplified, history becomes a political education, as well as a moral engine. Select Object Lessons in History would be such as these: namely, the Constitutions of some of the more primitive nations, beginning, for example, with the Hill tribes of India, and leading up by degrees to our own Constitution. As a select lesson the topic ' Revolution ' could be given, handled in the usual two forms, particular and general, or comparative. Some one Revolution, as the French, would supply a lesson in the particular or concrete ; and a comparative view of different revolutions would be given apart as the *general* exercise.

History is taught in the two alternative and mutually supplementing methods—a comprehensive sketch of Universal History, and a full detail of select periods. The Universal History would embody a Chronology, which is the chart of history, and with that the great leading events of the world. A somewhat fuller view might be given of Modern European history ; and a still fuller view of our own history. Outside of this would be set forth a minute sketch of certain epochs, say the turning epochs in our own history—the Norman Conquest, Magna Charta, the wars of England and Scotland, the Reformation, the Commonwealth. The least satisfactory compilations are those that are neither wide enough to give a general grasp of the world's history, nor minute enough to exhibit the historical forces at work.

Ancient History has been hitherto associated with classical studies, and thus introduced into a later stage of the pupils' course ; being scarcely at all mentioned, except by select episode, in primary instruction.

The teaching of history almost appears to defy Method. Any and every method would seem to apply, if we may judge by the variety of views that are entertained. The mistake is that the precise situation of the teacher is too little taken into account. He brings forward history, in the first instance, not for its own sake, but to help him in other branches. Thus history shares the fate common to many compositions, including the Bible itself. It is used simply for learning to read and spell. It is the vehicle for some of the first lessons in right and wrong, good and evil. It serves the use assigned to it by Goethe, to inspire enthusiasm, which might receive a wide interpretation, and imply the passions generally. In all this, there is scarcely anything of the distinctive functions of historical composition. The rules of method in these exercises are to be sought in other connections.

History proper starts with the idea of a nation or nations, and therefore supposes some knowledge of the structure of political society. This, I have remarked, must be the theme of distinctive lessons, which will be all the easier, according to the advancement made ín Geography, whose finale, Political Geography, is the true opening of history. When we understand what a nation is, we are prepared to follow its movements, changes, progress, and these are what history has to record. The narrative of events, to be of any value, should proceed upon the understood characteristics of the nation, and should throw back light upon these, like the mutual play of structure and function in Biology.

It would take a much higher acquaintance with political science than teachers usually possess, to conduct historical teaching upon any such method of mutual dependence. The pupils can reach the heights only by very slow steps, and it is entirely at the teacher's discretion, what explanations he should introduce at different stages, or at what stage he should take this view of history at all. Leaving the very earliest teaching use of history, there is the intermediate use—as animated chronology, or the succession of the great leading events of the world, in their order of date ; which is the setting of all superadded knowledge. The teacher can easily rise above the hard enumeration of dates, and the bare mention of dynasties, the rise and fall of nations, and other epochs,—to a more exciting narrative of the circumstances attending the more momentous events. The four ancient monarchies, the fall of the Roman Empire, the rise of the modern nationalities, the Crusades, the capture of Constantinople, the Thirty Years' War, the Wars of the French Revolution, the advance of European Colonization,—have to receive their fixed dates, so as to make a chronology, and to this may be added some very general statements of the agencies that operated such vast changes. Historic record is exceedingly elastic; it can be given justly on many scales : a very comprehensive view of the rise of the Greek power in the ancient world, and of its subversion by Rome, may assign the reasons or causes in an intelligible, interesting and correct, although greatly abridged, account. To be able offhand to vary the scale of the record is one of the arts of the teacher. He works upon a compilation, gene·rally too minute for his purpose, and he must know how to contract it; on other occasions, he may throw in the intermediate particulars, so as to enlarge the scale. This is only repeating an operation needed in Geography, as well as in other things.

The highest form of history is represented in the great works on the subject, ancient and modern. In these the structure of Political Institutions is more or less fully set forth, and the events treated on the deepest laws of political cause and effect. This kind of history is in alliance with the most advanced Sociology or Political Philosophy of the time ; and, as it is too extensive in its scope to be made a branch of a regular teaching, except by selection, the philosophy must count for more than the narrative. In fact, in the higher teaching of schools and colleges, history should be reduced to a science, and the narratives merely cited in exemplification of the principles. It is impossible to treat of all history; and epitomes or compendiums, upon the plan supposed in the earlier stage, would give no satisfaction; the only principle of selection is the exemplification of the theories of political cause and effect. The historical details, as given in the exhaustive histories of Greece and Rome, in the ancient world, and of the great nations

that make up the modern world, are overtaken afterwards by private reading. To repeat these in Lectures would be a waste of time, and must be at least
very fragmentary; and unless the portions actually comprised are chosen by accident and caprice, the ruling
consideration must be historical causation, given under
some system not at all difficult to shape in the present
state of political science. The questions even now asked
in. connection with history, at the higher Examinations,
prove that there is no difficulty in conceiving a suitable
culture for the most advanced stages of education in
this department.

Universal History having grown to interminable dimensions, it passes the compass of any single mind, and
would be a useless acquisition. Like many other subjects
in the present state of knowledge, including the chief
sciences—Mathematics, Physics, Chemistry, Biology—it
has to be taught upon some principle of selection. This
is not difficult to state. What has already been prescribed as representing the early stage of history
lessons, is supposed first of all. Next is the theory of
Political Society, and a comparative view of the leading
Institutions. Lastly, there should be some compact
body of Principles embracing the historic forces, with
their exemplifications in special portions of actual
history. More than one period would be desirable;
and the ancient and modern world should both be
represented.

The fact that history presents no difficulty to minds
of ordinary education and experience, and is, moreover, .
an interesting form of literature, is a sufficient reason for
not spending much time upon it in the curriculum of

school or college. When there is any doubt, we may settle the matter by leaving it out.

A very searching historical enquiry into modern events brings out such a variety of opinions in practical politics and still more in religion, as to make an obstacle to the introduction of the subject into the higher schools and colleges. This difficulty is felt in Germany, where professors are more outspoken than in England; it also occurs in connection with the Irish Roman Catholics in the Queen's Colleges. A history of the Reformation could scarcely be thorough, if it offended neither Protestants nor Roman Catholics; a history of the first centuries of the Christian era, if it dissatisfied nobody, would be worthless to everybody.

SCIENCE.

The Methods of teaching Science are as extensive and various as the field itself. They involve, in the highest degree, all the devices for the perspicuous communication of knowledge, as well as the more special devices of imparting generalized or abstract notions and truths. The teacher is usually supposed to have before him an exposition already shaped. He may of course modify the pre-existing exposition, and be a book to himself; but a convenient line may be drawn between the art of writing an expository work of science, and the art of bringing home the truths in actual teaching.

The methods of inculcating the abstract idea were incidentally sketched at an earlier stage. These methods are complete, as far as concerns the central fact of science —the generality or abstraction; only they do not include all that pertains to the teaching. Next to the exposition of a single abstract notion or principle, is the setting forth of demonstrative reasoning in chains of abstractions; a process that has its own separate difficulties. The requirement here is not a new expository method, but the careful employment of the ordinary forms of perspicuous language; to which is to be added the making sure that the links of the chain are each made secure.

Arithmetic.

The method for Arithmetical teaching is perhaps the best understood of any of the methods concerned with elementary studies. To illustrate Number by examples in the concrete, and to show the reasons of the rules by means of these examples, are the substance of the modern method, as opposed to the older practice of prescribing the tables and the rules, to be committed to memory, and carried into operation as the pupils best might.

Much is involved in the first attempts to work upon number. The distinction between one number and another is shown to the eye by concrete groups of various things; the identity of number appearing under disparity of materials and of grouping: ideas are thus acquired of unity, of two, three, &c., up to ten in a row. Difference or contrast is made use of, as well as agreement; five is placed by the side of four and of six. At the outset small tangible objects are used—balls, pebbles, coins, apples; then larger objects, as chairs, and pictures on a wall. Finally, dots, or short lines, or some other plain marks, are the representative examples to be deposited in the mind as the nearest approach to the abstract idea.

The conception of Number is not complete till it carry with it the ideas of more and less, of adding and taking away, and of the converting of one number into another by these means. More and less stands contrasted with the fundamental notion of equals, which also comes to the front in the first manipulation with numbers. Sameness in difference is exemplified in the notion of each number, obtained by comparing the concrete examples; and equality first conceived by coincidence of lengths, is transferred to number, by numerical coincidence in differently arranged groups; as when nine is set forth in one row, and also in three rows.

On the basis of the preliminary exercises with numbers in the concrete, the decimal system is reached, and with it the methods of adding and subtracting ; all which can be made quite intelligible and rational, as the precursor of the exercises to be worked. The sums of the simple numbers, having first been exemplified, have to be committed to memory; and this is the commencement of the business of computation, and of all the severe part of the subject. It being the essence of the abstracting operations to enable us to leap to conclusions, without going through all the intermediate steps, the memory has to receive with firmness and precision all that is included in the addition and multiplication tables ; and the test of aptitude for the subject is the readiness to come under this discipline. It is a kind of memory that in all probability depends on a certain maturity or advancement of the brain ; so that no amount of concrete illustration will force it on before its time. On the old system, the pupil commenced arithmetic when able to imbibe the tables and to work sums without any preliminary explanations of number; the ability arising in due course by the growth of the brain, and not depending on any aid from the teacher.

I am not aware of any special device for lightening this part of the process of arithmetical training. The general arts of teaching are of avail here as elsewhere— the apportioning of the lessons in suitable amount, the graduated exercises, the unbroken application, the patience and encouragement of the teacher. It is plain, however, that the multiplication table is a grand effort of the special memory for symbols and their combinations, and the labour is not to be extenuated in any way. The associations must be formed so as to operate automatically, that is, without thinking, enquiring, or reasoning; and for this we must trust to the unaided adhesiveness due to mechanical iteration. It is not unimportant to have gone so far into the rationale of the process as to be able to work out any one product deductively; and it might be a certain relief in the work of committing the table, to select a few of the products for determination by manipulating the factors—four sixes equal two tens and four, seven twelves equal eight tens and four. Another collateral exercise would be to call attention to each column as a steady addition of one number—twice six, three times six, and so on; which is the point where adding passes into multiplying. These explanations are useful in themselves, as contributing to the science of the subject; and they are a slight aid to the memory; we are not so apt to forget that four times six is twenty-four after having formed the twenty-four from the sixes. Still, I apprehend that the cementing of the requisite associations of the one hundred and forty-four products must be mainly an affair of symbolical memory, the result of immense iteration, and not to be entered on until a suitable age.

While this complete and self-sufficing association is the groundwork of the process of multiplication, which enters into all the higher operations, there are various points in the actual exercises, where the intelligent conception of numbers comes in aid; as in the placing of the multiplier below the multiplicand, and the arrangement of the lines of the successive products. For these matters, a knowledge of the reasons is very serviceable. The same applies to Fractions; in them the reasons assist the mind in observing the rules, which are not so easily held in the unmeaning shape as are the addition and multiplication tables. Still more does the knowledge of reasons apply to the Rule of Three, which can hardly be applied under any mode of stating it that does not assign the explanation. Hence, this is justly counted the *pons asinorum* of Arithmetic; it is the place where mere rote acquirement is sure to break down. So long as the questions are given in a regular form, the unmeaning rule may be enough; but as against distorted arrangements, it is powerless. In the usual applications to computing Interest, the hackneyed rule suffices only for the easiest cases.

It is thus apparent that, while many of the links of arithmetical operations are blind unmeaning symbolical associations, which are possible at a certain age, which may be called the dawning of the Age of Abstract Reason, because it is the epoch when the mind can betake itself to symbolical and representative signs, and think and operate through their instrumentality,—yet there runs through the subject a necessity of perceiving the grounds and connections of the various operations, and unless this perception is arrived at, there will be incessant halting. Nevertheless, when the proper routine is once learnt for all the recurring cases, the only thing wanted is facility in the cardinal operations, the result of the symbolical memory.

The full bearings of Arithmetic, as a science, cannot be seen until the pupil has made some way in the higher branches of Mathematics; and they are never completely known, except to the few that attain the conception of the highest scientific or logical method. In the lower stage of school training, ease and accuracy in calculation, extended to the ordinary compass of arithmetical problems, must be chiefly looked to. The persistent practice of years should bring about this result; while *rapidity* is attained by special drill in mental arithmetic.

There is an important principle of economy in Education that applies to Arithmetic, but not to it alone; that is, the utilizing of the questions or exercises, by making them the medium of useful information. Instead of giving unmeaning numbers to add, subtract, multiply, and so on, we might, after the more preliminary instances, make every question contain some important numerical data relating to the facts of nature, or the conventional usages of life; anticipating, as far as may be, the future exigencies of the pupils in their station in life. Not that they should be asked to commit these data to memory, or be twitted for not having attended to them, but that in those moments when attention is not engrossed with the difficulties of the purely arithmetical work, it may chance to fix upon the numbers given in the question, and thereby impress these on the memory

For example, the leading dates in chronology might be embodied in a variety of questions. A few sums and differences derived from the reigns of the English sovereigns, would be a collateral aid in stamping these on the memory; and might be the more effectual that it is not given as the essential stress of the exercise. Such simple examples in subtraction as how many years have elapsed since the Conquest, since the death of Charles I., since the Union of England and Scotland, the dates being either given in the question, or assumed to have been otherwise given--would help to impress these on the memory.

In a similar way, important Geographical numbers could be stamped on the recollection by being manipulated in a variety of questions. The dimensions, area, and population of the three kingdoms, the proportion of cultivated and uncultivated land, the population of the largest cities, the productions, trade, taxation of the country,—all which become the subject of reference and the groundwork of reasonings in politics,—could receive an increased hold on the mind by their iteration in the Arithmetical sums.

The common weights and measures should be familiar to everyone; and these might be so wrapped up in exercises, that the pupil could not avoid taking note of them. The mere act of writing them a number of times on the slate, with a view to solving questions, would render it· almost impossible to escape being struck by them. A most valuable datum in the ordinary contingencies of life is the relation of weight to bulk, given through the medium of water. A cubic foot of water weighs $62\frac{1}{2}$ lbs, and a gallon weighs 10 lbs.; these are data that no mind

should be without.   If a few leading specific gravities —
cork, wood (of some of the commoner kinds), building
stone, iron, lead, gold—were added, there would be the
means of readily arriving at many interesting facts.

Frequent reference might be made to foreign moneys
and scales of weights and measures, as of almost universal
interest; and especially to the decimal system of foreign
countries.   All this could be done in questions.

Next, I might cite the scales of the thermometer.
For want of knowing these, the statements of tempera-
ture are, in the majority of cases, unintelligible: the
Centigrade and Reaumur being now more in use than
Fahrenheit.

The comparative strength in alcohol of spirits, malt
liquors, and wines might be incidentally remembered by
being involved in exercises of computation.

We might range over the various sciences for interesting data.   Thus, in Astronomy, such leading numbers as
the sun's distance, and magnitude, the moon's distance,
the distances of the greater planets from the sun, the
periods of revolution, and rotation— could be chosen, as
not unlikely to make an impression through their incidental use in questions.   Such is the so-called perversity
of human nature, that the mind would often take a delight in dwelling upon these casual figures, because to
remember them was *not* a part of the task.   And further,
by a general law of the mind, if a question for some
reason or other has engaged the attention in an unusual
degree, the memory will receive the indelible stamp of
all its parts and accompaniments.

The Higher Mathematics.

The Methods in Geometry, Algebra, and the Higher Mathematics, are the methods for impressing abstract and symbolical notions and principles. The understanding must now accompany the work throughout; the stage of routine manipulation, worked up to automatic dexterity, is left behind. To a certain extent, the mechanical processes may enter into Algebra; the pupil may receive certain instructions, and, without understanding the reasons, perform the simpler operations of adding subtracting, multiplying, as in Arithmetic, but in the resolution of equations, the principles must be understood.

To advance at a moderate, steady, pace, to see each step well familiarized, before entering on the next,—are the rules of all difficult acquisition, from the beginning of time till the end. The earlier parts of such subjects as Geometry and Algebra need the longest iteration: the progress should be at an accelerating rate. The higher Mathematics should not be commenced with immature or incapable minds.

The fundamental axioms of Mathematics, including Arithmetic, are brought forward exclusively in connection with Geometry, which has always been the purest type of a demonstrative science. This makes Geometry now, what it was to Plato, the portal of the sciences. The scheme of formal demonstration, proceeding from Definitions, Axioms, and Postulates, is first unfolded in Euclid; hardly anything corresponding is found in the usual modes of commencing Arithmetic

21

and Algebra. No one knows Geometry, in the proper
scientific way, without comprehending the precise drift
of all these preparatory elements, as well as the nature
of consecutive demonstration. But there is a concrete
handling of Geometry precisely analogous to the Pestalozzi system for Arithmetic, and having the same effect.
It familiarizes the mind with the figures or diagrams,
enables sides and angles to be understood, and gives a
mode of experimental proof of some of the leading
theorems that is really conclusive in itself, although not
the sort of proof that belongs to the science. That the
sum of the three angles of a triangle is two right angles,
can be proved in the concrete; just as we can prove that
six times four is twenty-four. It would be a mistake,
however, to suppose that the experimental proof of propositions by cutting and folding cards is either Geometry,
or a preparation for entering on the march of Euclid, or
of any other system of Geometry conceived in the scientific form. When we come to the real business of Geometry, we have quite another sort of work before us; we
are refused appeals to the senses or the concrete, and
must establish each property as a consequence of some
previous property, starting first of all from the definitions and axioms, which are to be conceived as purely
representative abstractions. The serious work of the
teacher lies in following this plan, and in using his concrete instances only in aid of the abstractions as they
are given in the definitions. 'A line is length without breadth.' Examples of lines in the concrete may
be given with this definition, but what the pupil must
learn to understand (with no small difficulty) is, that
every concrete line is false to the definition; and that

the mental operation to be performed is thinking of the length, and neglecting, or leaving out of account, the breadth.   Next the *straight line* is taken up in a fashion that leaves the concrete far behind.   No doubt, a little concrete illustration is useful as a help to the definition—'lines cannot coincide in two points without coinciding altogether ;' but the notion must thenceforth be grasped as an abstraction, and conjoined with other abstractions in chains of demonstration.   So with the other definitions.   So also with the axioms : a few concrete examples are provided at the outset, and their support is thenceforth withdrawn ; the mind must hold by the abstract conceptions, as embodied partly in diagrams, and partly in general language ; and must be ready to draw inferences from clusters of propositions given in this naked form.   The concrete preparation soon exhausts its efficacy ; and the pupil has to depend upon the power to retain and to accumulate abstractions for the purposes of the work in hand.

The aid to be afforded by the teacher in mastering the demonstrations of Geometry, consists chiefly in making the essential steps prominent among the long-winded repetitions of subordinate matters.   The propositions as given in Euclid could be simplified by giving a larger type to the main statements ; and the living voice of the teacher can still further contribute to put the stress of attention where it is most required, and withdraw it from the tedious repetitions.

Algebra is better learnt *after* Geometry, inasmuch as it works in part by demonstration or deduction from principles, for which by far the best commencement is Geometry.   It has its own speciality, which consists in wrapping up the problems more completely in symbols, so that the inferences have to depend upon the validity of the symbolic representations and processes. The symbolic processes should be justified by explanations and demonstrations at the outset ; and the pupil should fully comprehend these. In point of fact, most pupils take all that upon trust ; the results being always right, and being easily verified, they go on the principle that 'all's well that ends well.' That 'minus multiplied by minus is plus,' is proved by never leading us to a wrong conclusion.

It is the province of the accomplished mathematician to provide the best possible simplifications of the difficulties that cloud the higher mathematics. How to embody the actual problems in mathematical language,—for example, the problems of motion in the scheme of differential co-efficients,—is a standing embarrassment, not to be met by any of the arts of ordinary tuition.

Mathematics is, in respect of teaching method, a sufficient type of the abstract and deductive sciences. All the subsequent sciences, in the fundamental group —Physics, Chemistry, Biology, Psychology—have their severe and abstract side, although with a growing dependence on the concrete. He that could teach Mathematics well, would not be a bad teacher, in any one of the rest, unless by the accident of total inaptitude for experimental illustration ; while the mere experimentalist is most likely to fall into the error of missing the essential condition of science as reasoned truth ; not to speak of the danger of making the instruction an affair of sensation, glitter, or pyrotechnic show.

An INDUCTIVE SCIENCE, as Experimental Physics, Chemistry, and Biology, is still a Science ; there are general principles and particular instances.   What is absent is the long consecutive chain of demonstration, requiring a sustained mastery of a whole series of propositions ; but this labour is replaced by others.   The laws of the Inductive sciences are liable to come under many qualifications and conditions ; instead of a single well-marked predicate, there is a complex and conditional predication ; and if we have not such a long course to run, we have another kind of mental tone, in the shape of manifold and distracting statements.

It is of great importance to reiterate, in connection with the general or fundamental sciences, that although many of their truths can be brought forward usefully, as object lessons, they will certainly not be retained in the mind with any degree of fixity or precision, unless they are finally grasped in their proper places in the mother sciences.   It is well to have interesting facts of heat, or of atmospheric pressure, exhibited in desultory fashion, at an age when Physical science could not be taught in its proper character ; but until the various facts are seen in their scientific connections, they will remain hazy and precarious.   Even a master of exposition, like Huxley, could not effectively impress, although he could clothe with interest, the truths of Biology, or Geology, in the form of an isolated address.

Natural History.

The Natural History Sciences are typified, and chiefly made up, by Mineralogy, Botany, and Zoology.   The Methods of teaching these are not difficult to assign, although there are some things that serve to complicate them. It is understood that they repeat facts, notions already obtained in the General Sciences, and that they are occupied with the arrangement, classification, and description of vast numbers of individual objects.

Any of these sciences, and particularly the two last, would swamp and overwhelm the strongest memory, and the details would be unprofitable when lodged there. The teacher has to hit upon a principle of selection that will guide him in making the most of a limited amount of time.

Take Mineralogy. There is here, as in the others, a *general* and a *special* department. The general department states fully, and in order, all the successive pro-perties of minerals at large—Mathematical (crystalline forms), Physical, Chemical—and adduces individual minerals as exemplifying the several properties, in all their various modes and degrees. This part the student is expected thoroughly to master; and after being familiar with the related mother sciences, can easily do so. It is, however, the smallest portion, as regards extent. The special department contains the classification of all existing minerals, and the enumeration and full descrip-tion of each species in its place. This cannot be com-prehended by any single mind. The lecturer is able to overtake the scheme of classification, with its divisions and subdivisions, and can do nothing further but choose exemplary species, for full description. A certain num-ber of species there are that embrace substances of such leading importance in the economy of nature that every instructed person might desire to know all about them; such are silica, alumina, lime, sulphur, the leading metals, and their more notable combinations. Others, as the gems, have the interest of beauty and rarity. So, everyone would wish to know something of the bodies that come to us from wandering over the realms of space. But of two or three thousand species, no teaching in an ordinary course could embrace more than forty or fifty, and of these the memory might hold in minute fulness, fifteen or twenty, and retain a vaguer recollection of the rest. The knowledge set forth, in the first or general department, might be retained with considerable firmness, and would include a great amount of *specific* information, in a very favourable form, namely, the enumeration of individuals as exemplifying general properties—cubical crystallization, highest degrees of hardness, magnetism, &c.

The arrangement is analogous for Botany. General Botany precedes, next come the Principles of Classification, and lastly the detail of Species, which is the interminable portion of the subject. There is a much shorter course, that is, recreative Botany, which teaches enough for the determination of wild plants, as encountered in their native localities. For this form of the science are compiled the *Floras* of countries and districts, which somewhat disguise the proper science of Botany; being still more remotely connected with Plant Biology, which deals with the processes of plant life, including the wide subject of fertilization.

Zoology is well understood to be a vaster subject than either of the others, whether as to the number of its objects, or as to their complication. Moreover, it is traversed by the special Anatomy of the Human Subject, by which the highest of Zoological species is taken out

of the classification and treated in isolation and independence, and with a degree of fulness that is accorded
to no other species. It is also to the human subject
that the laws of the animal mechanism and vital processes are attached, so that the mother science of Biology
in its Animal department is almost entirely concerned
with Man; other animals being used as subsidiary elu-
cidations. There is thus a play of cross purposes between Human Anatomy and Physiology, Comparative
Anatomy and Physiology, and Zoology proper; and as
they make three vast subjects, no one can overtake more
than one, with a small portion of the others; and the
selection of what is most valuable for a general education
is yet a desideratum. Biology as Human Physiology,
enlarged by comparative references to the animals generally, should enter into a complete scientific training.
Now the human subject, if anatomically studied, which
is hardly possible except in a medical school, is so far
a key to the Vertebrates generally, and the Mammalia
especially, that it contains all their parts and something
more; yet there remain very considerable differences;
and Zoology is still an arduous and extensive study,
which must be reduced by selection, until even whole
Classes, not to speak of Natural Orders, Genera and
Species, are left unrepresented in a tolerably extended
course. Still, the groundwork may be laid for following
out the subject, which is all that teaching can do, or
should attempt, for many of the most fruitful regions of
knowledge.

Geology has a sphere of its own, although involving
references to all the three foregoing sciences. It is a
science of detail but not to the extent of the others;

and the physical processes are more thought of than in them; in which respect, it nearly resembles Meteorology, an applied department of Physics. Geology could be understood and followed on a basis of Natural Philosophy or Physics, with mere 'object lessons' on minerals, plants and animals.

PRACTICAL TEACHING.

With reference to the Experimental Sciences of Physics, Chemistry, Biology, and the Natural History group, it is now frequently urged that the teaching should be practical : meaning, not merely, that the teacher should present to the pupils actual experiments and specimens, but that the pupils should manipulate with their own hands. Professor Huxley seems to hold that Zoology cannot be learnt with any degree of sufficiency unless the student practise dissection.

In support of this position, there are strong reasons. In the first place, the impression made on the mind by the actual objects, as seen, handled, and operated upon, is far beyond the efficacy of words, or description. And not only is it greater, but it is more faithful to the fact. While diagrams have a special value in bringing out links of connection that are disguised in the actual objects, they can never show the things exactly as they appear to our senses ; and this full and precise conception of actuality is the most desirable form of knowledge ; it is truth, the whole truth, and nothing but the truth. Moreover, it enables the student to exercise a free and independent judgment upon the dicta of the teacher.

Whether the power and habit of experimental manipulation should be acquired for its own sake, depends on what further use is likely to be made of it. In the Zoology courses in the 'School of Mines,' there are schoolmasters' classes, where dissecting is practised, and is useful; but we cannot contend that very valuable instruction may not be imparted by merely showing dissected and prepared specimens, although the pupil has no hand in the work. So, in Experimental Physics, a good knowledge may be obtained from a course that shows all needful experiments, without the actual participation by the pupils themselves. To make an experiment succeed, many delicate precautions and fine manipulations may be wanted; some of these precautions implicate matters of knowledge that are not perhaps conveyed to the minds of the learners, while they are very strongly impressed on the mind of the experimenter. As regards the mere manual skill, that cannot be called a part of scientific information or discipline, while to acquire it needs time and attention. The Laboratory teaching in Physics (a recent innovation), like Laboratory teaching in Chemistry, is a good introduction to various scientific avocations, as Engineering, Machinery, and Manufactures; it cannot be regarded as essential to the general course of scientific study, and would be too dearly bought at the cost of marring some other department of science. More especially, if training in the higher intellectual operations of the mind is the object of view, would it be a disproportion to give up a large share of time to practical working.

What may be said in favour of practical teaching for a *general* training is, that the arts, devices, precautions connected with exact observation, would be brought

home by a course of study in *some one* of the experimental or observational sciences. Practical experience
in a single subject would be enough ; and the interest of
the work would go far to repay the devotion. It is plain
that in none of the experimental subjects could anyone
be an adept, an expert, or an authority, apart from the
practical study ; but to carry the information and the
training forward into other departments, a knowledge
obtained without this help may suffice. We need not
be workers in Physiology to apply its results to the
physical accompaniments of the Mind.

ORAL TEACHING AND TEXT-BOOKS.

In primary instruction, and, to a certain extent, in
the secondary higher instruction, text-books are made
use of as the means of communicating knowledge. They
are very variously employed. Sometimes the teacher
himself orally delivers the whole substance of the lesson, referring to the text-book as a further aid. Sometimes he selects portions for oral exposition, thus awakening the pupils' minds to what is to be done, and leaving
the rest to their own exertions. Lastly, he may do nothing
at all, but exact, in the form of lessons, an account of
what is in the book, giving corrections and explanations
according as these are found to be necessary. The first
method very nearly approaches to independent lecturing,
the text-book being an adjunct and support. A combination of the lecture and the text-book, when they are
made to harmonize well, is an effective mode of carrying
on instruction both in the lower and in the higher grades.
The text-book does not supersede the lectures, but only

supplies gaps. If there is no text-book, provision has to be made for taking full notes, and the lecturer must advance slowly, and be careful to dictate or post up the heads and leading principles.

Lecturing, that is, oral teaching, with or without text-book, has the very great advantage of the living voice, aided by the sympathy of numbers ; and is indispensable in school-teaching. Young pupils have much difficulty in guessing out for themselves the meaning of a compactly-worded handbook; to set them to work at this, as an evening task, is a kind of fatigue drill. The generality are found at fault when the class is examined; a few may succeed, and the others get the benefit of the rehearsal, with the comments of the master, and in that way learn all that they do learn. If a lesson, after being heard in this way, were to be again prescribed, the getting up might be extended to the whole class.

A task may be of a kind to dispense with preliminary explanation, as in learning a string of words, or a verbatim statement. Even then, it is well that the teacher should first recite it to the pupils ; his doing so once will go farther to fix it in the memory than their going over it by themselves six times. There is no harm, but good, in exacting a certain amount of independent preparation, especially with older pupils, but the teacher's first recitation, and the final iteration during the lessons, are the principal instrumentality whereby the lesson is fixed in the memory; the learner's own studies are the smallest contribution to the effect.

When there are difficulties to be encountered, the previous explanations of the master are indispensable. One species of difficulty deserves mention, as serious in itself, and of common occurrence. A passage in the text-book may be prescribed, not to be got by heart, but to be understood, and repeated *in substance*. In historical narration, in a geographical picture, in a natural history description, and in scientific explanation, this position is created ; and it is a severe ordeal alike to the text-book, to the pupil, and to the master. To divine what are essential or leading points in a subject, and what are the accessory or subordinate details, requires a very mature judgment ; while, if they are separated by typography, in the text-book, the lowest degree of judgment is dispensed with, unless the exercise be to show the bearing of the one on the other. But for a pupil to discover what shall appear the leading points to the master, is to be something more than a pupil. It is in this situation, that the teacher should indicate what the points are, and should awaken the minds of pupils to the difference between these and the subordinate details.

The easiest case is when there is a principle or rule, with a host of examples to choose from ; all that is necessary is to prescribe the rule with a choice of examples. The exercise is not one of memory but of comprehension.

It is bad policy to prescribe lessons of excessive length, expecting only a part to be performed. If, for the sake of the better pupils, the lesson should exceed what the average can perform, the minimum should be a defined portion, to be exacted of everyone. The impossibility of bringing every pupil in a class to book, on every occasion, is in itself a standing temptation to run · the blockade ; but when the quantity prescribed is beyond what can be reasonably required, the do-nothing habit receives positive encouragement.

EXAMINATIONS GENERALLY.

Examinations are a part of the means of making pupils put in practice what is taught them. The first step is to show or tell them something, the next is to make them do it themselves. As regards information imparted, the exercise consists in making them rehearse it, to show that it has taken hold of the memory. If it be a matter of reasoning, they must do something to show that they comprehend the reasons; for which case, a test must be devised to distinguish between repeating verbatim the words of a reason, and understanding its bearing. To this every good teacher or examiner is competent; the means are not far to seek. Of all the evasions coming under the designation 'cram,' the substitution of memory for understanding is the easiest to unmask.

The most singular abuse of the process of examination is seen in the time-honoured composition called the Catechism. Although most identified with religious teaching, the catechetical form has extended into all sorts of subjects. The point of it is to give the teacher the words of the questions to be put to the pupils, while they are to repeat the words of the answer, and so fulfil their part, as in a liturgy. It is true that good teachers nowadays superadd a cross-examination, but this is to innovate on the very idea of the catechism, and shows that the time has come for superseding it. It has been for ages the vehicle of a purely mechanical teaching.

There is a reason for occasionally appending a series of questions to passages intended for conveying information, namely, to call attention to leading points, and to guide the pupils in preparing their task, as well as to

assist the master. Such questions do not lend themselves to mechanical teaching, but may do very much
the reverse. No doubt each master should be able to
put them for himself, but the giving of them beforehand
is an advantage, by setting something for the pupils
to do.

The conducting of Examinations was originally
viewed as a part of the teaching; and the point considered was how much and what kind of examination
should go along with *vivâ voce* lecturing; and under
what circumstances it was justifiable to prelect without
examining at all, as in the German Universities and
elsewhere. The Examinations at the close of the course
or curriculum, were merely questions analogous to those
put during the teaching, to show whether the pupils
retained in their memory to the last what they seemed
to imbibe from day to day. Upon such examinations
certificates and prizes were awarded and degrees con-
ferred.

The present system of bestowing important State
offices by Competitive Examination has given a new importance to the methods of conducting them; and the
whole subject is undergoing fresh and rigorous scrutiny.
But if we are to go to the root of this matter, we need
to consider three things in succession. I. What are the
subjects that constitute the best groundwork of intellectual power? II. How are these subjects to be taught?
III. How are they to be tested in Examinations? The
first of these topics is very far from being yet settled;
and I have devoted a considerable space to elucidate it
according to my best judgment. The second topic is the
question—How to teach? which still more largely occu-

pies the present work. The third topic I refrain from entering upon. A very thorough discussion has been recently given to it, in a work 'On the Action of Examinations considered as a means of Selection,' by Mr. Henry Latham, of Trinity Hall, Cambridge.

CHAPTER IX.

THE MOTHER TONGUE.

THE Methods for acquiring Languages contain certain peculiarities that render desirable a separate discussion. The questions respecting the education in the mother tongue are numerous and highly involved. As regards Foreign Languages, there remain to be adjusted, not merely the modes of acquisition, but the disputes as to *value.*

There is an easily-conceivable state of things, that would dispense entirely with school instruction in the mother tongue. If the child were surrounded only by those that spoke correctly and well; if it had ample opportunities of coming into contact with all the highest resources of the language, and of being imbued with the most tasteful usages ;—then the education in the mother tongue would be perfect through unavoidable imitation; there would be no other teaching than there is of the provincialisms of dialect and brogue. Something approaching to this occurs in the better classes of society with ourselves; and it is the whole case among nations that have but one set of expressions for all subjects, and for all persons.

It is in so far as our actual position is different, that we need express teaching in the native tongue. Yet allowance has always to be made for the unavoidable and incidental instruction. It is an error to repeat at school what is learnt at home; and a still commoner mistake to occupy school time with what is sure to be . learnt in the great school of the world. Every hour spent in society is fraught with lessons in language. All knowledge acquired by the help of others comes to us in its language garb. A wide course of attainments in the various departments of knowledge is inseparable from a culture in the means of expression.

The primary school has to fight against the low standard of the home, in language as in other things. The other schools maintain the same contest; and the further contest with what is bad even in the speech of the educated ; including the mixture of tares and wheat in the field of general literature.

The initial and pervading difficulty in teaching language in general, and the mother tongue in particular, is due to the doubleness of the acquisition—the union of language and thought. Language is nothing without thoughts to express; and the attention is divided between the two factors, instead of being concentrated upon one to the neglect of the other. Moreover, the kind of thought to be expressed must needs affect the manner of expressing it, and must, therefore, be taken into the account. There are, however, many of the arrangements of language that are the same for every variety of subject: such are the proprieties of Grammar, and a certain number of the conditions of Rhetoric; these constitute the more purely language studies.

It has to be borne in mind that to teach language is not to teach knowledge, in the sense that we usually understand knowledge : as history, geography, science, the arts.   Neither is it to impart lofty or poetic sentiment or moral elevation.   In those matters, language is the vehicle or instrument ; but while we are *using* it for such purposes, we are not expressly teaching it.   True, to employ language for any purpose is an indirect and un-intentional way of giving instruction in language ; and a large part of our language education is gained in this form ; but such an effect is to be kept distinct from lan guage exercises properly so called.

The situation of carrying on a double subject is of frequent occurrence in our intellectual culture ; many branches of knowledge have a twofold aspect ; and it is easy to go wrong in dealing with this situation.   The fact never to be forgotten is that the human mind can attend to only one thing at a time, although it may shift the attention very rapidly, and thus overtake two or more things by turns.   In matters of education, however, where different subjects have to be mastered, or where numerous details have to be impressed on the memory, concentration on one exercise for a certain time is indis-pensable ; and, in those subjects that proceed on a double line, the attention should be sustained in one of the two directions, instead of flitting between both.

For the greater part of language instruction, this principle of distinct attention can be fully carried out. Yet there are a few situations where the language and the thought are incorporated in such a way that they cannot even be separately considered.   In Wit and Epigram, and in Figurative illustration, the ideas and the words equally constitute the vehicle for the thought to be ultimately conveyed.   On the other hand, a knowledge of things often assumes the guise of knowledge of language. What is an 'epidemic'? seems a verbal question, but in reality it is a demand for knowledge as to a natural fact or phenomenon.

CONDITIONS OF LANGUAGE ACQUISITION GENERALLY.

Language, as an acquisition, is seen in greatest purity in learning foreign languages. It is in these that we exemplify the process of adding word to word by verbal adhesiveness. The mother tongue involves largely the operation of associating names with things or thoughts. The laws of acquirement are not the same for the two cases.

In every form, language is a very serious draft upon the plastic power of the mind; and needs to be favoured by all those general conditions of retentiveness formerly set forth. The detailed associations that have to be constituted are exceedingly vast in amount; the mere number of vocables in any cultivated language counts by thousands, while many of them have plural meanings. To these must be superadded the special meanings of phrases, ideas, or combinations involving distinct acts of memory.

In the process of uniting word to word there is exemplified the purely verbal adhesiveness of the mind, embodied principally in the ear, for spoken language, and in the eye for written language; the voice being an adjunct to the ear, and the hand to the eye. This is the least favourable mode of learning language. To connect an English word with the corresponding word in French, in Latin, in Greek, does not bring into play the most powerful of the associating forces; it does not fall into the most adhesive lines, and it is not supported by the higher degrees of interest.

It is in associating names *at once with their objects* or meanings, that the acquisition of language proceeds most rapidly This is our situation in learning the mother tongue. We are in view of some object—a fire, a ball, a cat—which lays hold of our attention for the time; the name falling on the ear, is fused in the same act of attention and becomes speedily associated with the object. The more sensational a thing is, the sooner is the name incorporated: a flash of light, a sudden noise, a rapid motion, a fracture—if duly named, at the moment of occurrence, scarcely needs a second repetition.

In this operation of combining names with our actual experiences, we are powerfully aided by the emotional response, which follows on every impression of any considerable force. We feel prompted to some vocal exclamation, whenever we are moved, or excited, or made to attend to anything; and, if we are made to hear the verbal designation, we fall into that, as the mode of venting the emotional impulse. The child soon shows this tendency to call out the name of anything that arrests its attention—fire, Puss, Tommy; and the command of language is in this way greatly promoted

When learning a foreign language in the country where it is spoken, we are made aware of the difference between adding word to word, and connecting each name at once with the realities. In a French town, we see the word 'rue' put up at every street corner; in the shop windows we see the articles laid out with their several names attached. Riding on a rough road, after some great jolt, we hear a fellow-passenger exclaim, 'secousse,' and one utterance is enough to attach the name to the situation for ever; whereas, several repetitions of 'secousse' and 'shock' would be requisite to establish a durable bond between the two words.

In strict propriety, the same effect ought to arise through the circumstance, that the words in our own language should, when mentioned, carry our thoughts to their meaning; the word 'secousse' when explained by the English 'shock,' ought thereby to become connected with what 'shock' expresses, and an association be thus formed between the French word and the actual situation. It is obvious, however, that the result is apt to be but feebly and slowly attained through this means. The process is rendered more effectual by dwelling for a little on the meaning, or giving explanations that serve to bring the actuality strongly before the mind, thus: 'secousse'—'a shock, or jolt, as when riding on a rough road,' &c. Such an explanation given orally by the teacher, is likely to be sufficient to fix the word; it is necessarily less effective when read by the pupil from the dictionary.

This single example is enough to show how the power of apprehending meanings is essential to rapid progress in language; that is to say, the knowledge of things should always keep ahead of the knowledge of terms. To force on prematurely the knowledge of unfamiliar subjects, in order that a very young pupil may learn a hard language, as Latin, is working at the wrong end. If we are to read any author as a lingual exercise, it is desirable that we should previously understand his subject as a knowledge exercise; we are then in the proper position for acquiring the vocables and forms of

language employed by him for expressing that knowledge. To learn the Greek phraseology for Geometry, we
should first understand Geometry, by studying it through
our own language, and should then read Euclid in the
original   It is not too much to say that the best geometer would make the most rapid progress in the Greek;
even the superior verbal memory of the young, or of
those unusually gifted with verbal adhesiveness, would
not make up for an imperfect hold of the subject matter;
a good mathematician of fifty would probably finish ·
the task sooner than a half-taught youth of fifteen, with
the memory for words almost at the very best.

Here we see one of the weak points of the early
study of foreign languages, and especially of dead languages. The strong point is the freshness and force of
the memory, coupled with the inaptitude of the reason
for the higher kinds of knowledge. But unless the language is acquired with reference to the things actually
understood by the pupil, it will not take hold even of the
best memory. The difficulty is not usually perceived,
in consequence of the universality of the easy-narrative
form of composition—the omnipresent resource in the
early stages of instruction. As it is very desirable to
teach early the pronunciation of a foreign language, and
as this needs some degree of acquaintance with the language itself, it may be taught in the nursery as far as ·
the knowledge of a child goes, but no farther.

The disadvantage of combining language teaching
with the teaching of things is, as we shall see, inevitable
(although it may be mitigated) in the mother tongue,
but need not be repeated, as it often is, in foreign languages. In these, the subject-knowledge might always

be well ahead of the language study. The seeming objections to such a course may be met by a preponderance
of counter considerations.

Postponing for a little the full application of these
generalities to the mother tongue, I must advert to the
purest case of verbal acquisition—the adding of word to
word, without reference to meaning. Although the staple
of our language power should consist in associating
words with things, situations, or meanings, there is still
a considerable amount of the other exercise implicated
in our learning of language, and in our knowledge as
embodied in language. Take, for example, the details
of Grammar—the inflections of inflected words, and the
lists that accompany grammar rules and definitions.
Likewise, the sentence forms that constitute our earliest
utterances, and the more elaborate forms that we use
with mechanical readiness in our later stages, are almost
purely verbal associations. So are the passages committed to memory, at a time when meaning counts for
very little with us; and all the sayings that we remember
for their verbal point. Knowledge reduced to language
is remembered by us partly through the coherence of the
subject matter, and partly through the coherence of the
words employed. Moreover, the synonyms of language
are in great part grouped by our verbal memory; and
this extends beyond vocables to synonymous constructions and varied expressions for the same thing. Of
course, all that part of the retention of foreign languages
that is dependent on associating the mother name with
its foreign equivalent comes under the same head. The
extreme case is the acquirement of languages by the
philologist for purely philological comparisons.

These purely verbal associations, required in such immense numbers, must be regarded as a disagreeable necessity; it is only by accident, that they are interesting in themselves. Who can feel any charm in committing to memory the list of irregular verbs and their irregularities ? The conditions that regulate our progress in this department are those that reign in all the dryest parts of knowledge. The plastic forces must operate without depending on any stimulus of charm or liking for the work.

The recognized principles of economy in acquisition need to be sedulously attended to, wherever the elements are both multitudinous and unattractive. The work has to be divided into portions for each day's share ; there is a proper allowance of time, strength, and attention, during the plastic periods of the day; the tasks are repeated carefully to the master, and stamped with his *imprimatur* and approval. Rehearsals of foregone tasks are made at due intervals. The young are stimulated by small gifts. These are the chief devices for overcoming dulness in all lessons of detail.

As to the particular case of word-lessons, there are facilitating arts that the careful teacher does not overlook. Their mode of presentation to the mind, if by oral utterance, should be characterized by clear, distinct, and even pleasing tones ; if by writing or print, the characters should be plain, the lists symmetrically arranged. It is also a valuable aid to copy out carefully, important lists, schemes of declension and conjugation, and other technicalities. This is known to be one of the ways of stimulating attention, at least so long as copying is not a mere mechanical exercise.

Next to these devices for the judicious management of the plastic faculty, we have the arts special to verbal acquisition, including the schemes of technical memory.

It is a well known means of alleviating the burden of language details, to be able to detect latent similarities in the words to be associated, as in learning German, French and Latin. Many such similarities are open and apparent; others can be made apparent by slight transmutations, which the teacher or dictionary can point out. This is the philological aid to the study of foreign languages.

The technical memory proceeds upon various other arts. The learning of lists of words is facilitated by the ancient device of casting them into verse; the saving by this means must be pronounced considerable. The making up of an intelligent sentence would have a similarly good effect; if the meaning were at all interesting, this plan would excel the other.

Alphabetical arrangement renders a train of words much easier to acquire; the link of alphabetical sequence being a sure one, lends itself as a help to the memory of words. Unless there be some special reason against it, this arrangement should always be followed.

Another device still more peculiarly technical is the arranging of words in such a way that the meanings have some natural connection calculated to aid the memory. This is a modification of the topical memory of the ancient orators.

With a view to the firm and permanent association of one word with another word, there should be a moment of isolated and concentrated attention upon the two. This is gained in various ways. The turning up a

dictionary is the commonest way of isolating the attention; to a young, fresh memory, one dictionary reference
is usually enough, provided the words soon recur.   Another way is to hear the words deliberately pronounced
by the master in translating a passage. The Hamiltonian
method does not provide for the isolating of the attention upon single words. A better method seems to be
to prepare a series of exercises, each embodying two or
three (and no more) new words.

The mode of teaching languages in use in this country
usually throws upon the pupil the labour of finding out
the meaning of a passage, in the first instance, by the
help of the dictionary; this being corrected and impressed by the repetition under the master. There seems
no reason why this should not be combined with a little
of the other method, namely, for the master to expound
a passage fully, in the first instance, and then require
the pupils to reproduce it next day, by memory, aided,
if need be, by a reference to the dictionary when something has been forgotten. One portion of a prescribed
lesson might be given in this way, and the second portion left to the old method. No doubt the method of
prior exposition is best suited to scientific lessons, as
Geometry, but it is not wholly unsuitable to lessons of
pure memory.

I am waiving for the present all questions as to the
grammatical teaching of language, and am merely adducing situations where the joining of word to word is
the essential fact.

In an interesting lecture by Mr. Alexander J. Ellis,
on the Teaching of Language, I find great stress laid on
the statistical valuing of words according to their fre-

quency of recurrence. It is urged by Mr. Ellis that this relative frequency should determine the order of presentation of words in exercises. The plan is carried out by Mr. David Nasmith in reference both to English and to German. It supposes that the statistical frequency of words shall have been previously ascertained. The full carrying out of the principle would embrace, not only words, but phrases.

I am disposed to think that this principle might have a much wider application than to language. It would be well if we could forecast the probable frequency of the use of every acquisition whatever, so that we might choose by preference those that oftenest come into play, and, I may add, on the most important occasions. Such a criterion would attest the high value of the Experimental Sciences, such as Physics and Chemistry, the smaller but yet considerable value of Mineralogy and Botany, and the very small value of many things much more prominent in our existing educations than any of these. As regards languages, however, the principle must be somewhat qualified. Although one word in a language may not occur so often as another, the two may be equally essential in the long run ; and it is scarcely worth drawing more than a few grades of distinctions, beginning at the most indispensable words of all, taking next those that occur in the most ordinary speech, and so on ; leaving to the last the rarer and more technical or abstruse designations. Besides, the mere fact of frequency operates of itself ; we do not need an artificial scheme in order to bring forward, at an early stage, the recurring words ; we need only keep back for a time the words that are not frequent or essential. Yet if a word has to be learnt sooner or later, the period of learning it cannot signify much ; the bringing forward of rare words before such as are common, could not in the nature of things be carried to the length of an abuse.

The foregoing considerations belong to all language, and refer to the most characteristic situation of language acquirement. We shall now turn to the special situation of learning the mother tongue, after which we can resume with more advantage the question of foreign languages.

THE MOTHER TONGUE.

At all the stages of learning the mother tongue, the purely verbal exercises are more or less accompanied with the occupation of the mind upon things. In Grammar, the disengagement from things is at the greatest.

If we suppose the child to become acquainted, in the first instance, with a variety of objects, the imparting of the names is a welcome operation, and the mental fusion of each name and thing is rapidly brought about. According as the object named has been fully perceived, and is well marked out from other objects, the learning of the name is easy. If the objects are in any way interesting, if they rouse or excite attention, their names are eagerly embraced. All through life we show an avidity for knowing the names of persons, operations, places, circumstances, that awaken our regards. On the other hand, if objects are but languidly cared for, or if they are inconspicuous, or confused with other things, we are indifferent both to the things themselves and to their designations. In this case, the first step is to secure a proper impression of the matter to be named.

The more we enquire into the early teaching of language, the more shall we find it to be in great part the teaching of knowledge under difficulties. The child is soon brought into the situation of having to comprehend consecutive speech, many parts of which are devoid of meaning. But to explain the words that are blank to the mind, we have first to bring before the view things that have hitherto been entirely unknown. We have to communicate a knowledge lesson, supplemented by a verbal lesson, the first being by far the more serious of the two; indeed, the second is as nothing, after complete success in the first. If the teacher can but compass the knowledge difficulty, he does not need any extraordinary efforts or any refined methods for securing the adherence of the verbal expression. The faculties that are awake at the early period we are supposing, are more spontaneously adequate for learning names when once the things are conceived, than for mastering the conceptions themselves.

The real solution of the difficulty of teaching language at the first stages of intelligence, is the Object Lesson, or whatever we choose to call the beginnings of imparted knowledge.

The chief bearing of language on the situation is this, namely, that the use of language, in speech or in books, is the occasion for bringing forward the things to be taught and explained. The best form of introducing a fact would be its real occurrence, as when the child sees the evening star, and is there and then told something about it. But listening to talk, and book-reading, bring forward things without any reference to their actual presentation; and then some way of introducing them has to be found ; the task being in many instances premature and impossible.[1]

It is in the situation of fragmentary knowledge, that the verbal memory may take hold of language without understanding, of which enough has been said in a former connection (p. 205.) There is a kind of pre-occupation of the mind with terms, that acts as a spur to seek out the meanings by observing the occasions when they are used. The child may be familiarized with a name, as 'light,' for some time before grasping the sense. There is a sort of inductive process gone through, in singling out the true meaning from among the surrounding circumstances. All the inductive methods of Logic are applied by the child in connect-ing 'light' with its true meaning. With general names, there must be a generalizing operation, as in arriving at the meaning of 'round,' 'heavy,' 'cold,' 'motion.'

All this is but the language side of knowledge, and does not represent language culture as such, or what constitutes the domain of the language teacher. In order to frame a special language department, we must assume knowledge as at a stand-still, and consider the different ways, for better or for worse, of expressing any known fact, doctrine, or set of facts or doctrines. The knowledge teacher provides at least one mode of stating what he communicates, but he does not occupy himself

---

[1] The explanation of the names occurring in the reading lessons is a large part of the teacher's work ; and the best methods of conducting it deserve to be studied. It follows in the track of the Object Lesson as dis-cussed in the foregoing chapter, but has certain distinctive peculiarities. In the Appendix I supply some additional illustrations to bring out the more delicate precautions in giving Object Lessons in their ordinary forms, and advert also to the present topic.

with weighing the merits of all the different possible verbal statements of a piece of information. Division of labour is necessary to educate the whole man ; affording, on one side, breadth of knowledge, on the other side, sufficiency, and even luxury, of expression.

Let us assume at the outset, that the teacher devotes himself to following up, or if need be, to rectifying the work of the parent, in securing good articulation, and pronunciation, and a certain propriety of accent. Nay more, the correction of vulgarisms and provincial errors, might be attended to from the first day that the pupil enters the school; there is no need to wait for Grammar rules, to put a stop to the grosser errors of concord prevailing among the lower orders in this country. Although the delicacies of syntax cannot be given by the ear, the practice of the more educated households shows that children may be taught at once to say 'he *is*,' 'they *are*,' 'that (for those) sort of things,' and to make the proper distinctive applications of 'shall' and 'will,' 'may' and 'can.' [1]

The real division of labour between knowledge teaching and language teaching comes into prominence at the Grammar stage; we do not confound the teaching of Grammar with the teaching of things. It is one of

---

[1] As the teacher, in the National school, has to fight against the force of almost unanimous out-of-door usage, he might obtain support by the use of a little printed manual of the prevailing errors or vulgarisms in his district, which might be used long before the age of Grammar. Such manuals need not exceed twenty or thirty pages, and might be produced at the cost of a halfpenny. Their diffusion in the homes of the pupils would be a powerful aid to the influence of the master. They could be composed without formal rules, but with a sufficient variety of examples to show correct usage in all ordinary cases

the advantages of a grammatical course to make the distinction apparent, to give an occasion for imparting language lessons, pure and simple. At the previous stages the teacher is often in doubt whether his teaching relates to things or to language, or to both; the fact being, that he is always fluctuating between the two. The following remarks are intended to point out fully the difference of the two lessons.

It has already been noted that the explanation of newly-occurring terms is for the most part thing-knowledge. When the word 'slave' is presented for the first time, an explanation of the state of slavery is provided, whereby a new idea is imparted to the pupil. This is in no sense a word lesson, although the occurrence of a word is the occasion for teaching the thing. If the pupil has had prior experience of things, without knowing their names, to give the name is a language lesson: this situation is not so frequent as the other.

The first decided exemplification of language lessons on the great scale is the teaching of synonymous words. The best example of this is the perpetual passing to and fro between our two vocabularies — Saxon and Classical. The pupils bring with them the homely names for what they know, and the master translates these into the more dignified and accurate names; or in reading, he makes the learned terms intelligible by referring to the more familiar. The multiplication of synonyms may be carried farther, by adducing figurative and poetical, as well as scientific, equivalents. Suppose the name 'die,' or 'death'; the learned equivalent is 'mortality,' but the figurative equivalents, together with the phrases and circumlocutions, are very numerous—'loss of life,' 'eternal

23

sleep,' ' paying the debt to nature,' ' passing out of being,
' extinction,' ' separation of soul and body,' ' sinking into
the grave,' ' going to our long home.' Now in extending the list of these equivalents, the teacher is giving a
verbal or language lesson.

There is, nevertheless, a subtle reaction of the knowledge of things, even in this exercise. The Figurative
equivalents imply comparisons between the subject and
other subjects, and are used to elucidate or intensify the
meaning, provided they are themselves understood;
while, if they are not as yet understood, the teacher may
wish to make them so, and thereby enter upon an object
lesson. As a rule, however, such digressions should be
forborne; if the figure is already known, it will work its
effect, if not, it will do nothing for the present; while
the name may be associated with the other names as part
of the stock of synonyms. The phrase ' separation of
soul and body' is hardly intelligible to children; but a
good memory could learn it by rote as a designation of
death, while the meaning is as yet very faintly conceived.

The explanation of figures is not the only way that
a lesson in synonyms may pass into a knowledge lesson
It is the fact that synonyms are rarely identical; they
give different shades or degrees of meaning, or they
present a thing from different points of view, or they
are more or less vague or precise. Now any attempt
to point out these distinctions would take the teacher
into the knowledge sphere. ' Truth,' ' verity,' ' veracity,'
' consistency,' have a common meaning, with differences
that prevent their indiscriminate application. To point
out these differences is to give a lesson in the subject

matter and not in the expression. Such lessons are not to be entered upon at random.

It may be said with some plausibility that names should be learned at the time that we are informed of the exact ideas that they represent. And no doubt this is a sound principle, and is to be respected so far, that we should not purposely bring forward names whose meanings cannot be taken in at the same time. But as we cannot keep back words at pleasure, our only course is to let them be known with such vague or approximate signification as the pupil can readily imbibe, leaving their more delicate shades to be gathered by subsequent experience. The temporary result is a malaprop use of words. The permanent utility is a command of terms for the purposes of selection.

Let us turn now, by way of illustrative contrast, to the knowledge-master viewed as also a word-master. We have repeatedly seen that the proper order of acquirement is—Thing first, Name second; which, if rigidly carried out, makes the knowledge-teacher the sole language-teacher. The limitations have just been noticed, and will appear more fully at an after stage. In the meantime, we are able to show that even as regards the wealth of synonyms, the knowledge-master need not be much behindhand. It is not his purpose to exhaust all the conceivable ways of expressing every fact within his subject; but the necessities of explanation, when the matter is difficult, lead him to cite a considerable number of ways, while he is almost sure to use them with discrimination.

Thus, to take the force of Gravity. The teacher in explaining this force, must trust chiefly to exemplifying

it by familiar facts, as the descent of unsupported bodies.
But he does not neglect the reference to the various names
that have been used in connection with the force: –
'weight,' 'downward pressure,' 'falling to the ground,'
'attraction,' 'drawing together,' 'mutual pulling,' 'de-
flection from a straight course.'

Some of these names are already associated in the
hearer's mind with the working of gravity, and others
will at some future time be associated, and are therefore
advantageously planted in the memory. The language
teacher could hardly go farther, in the direction of mere
synonyms, at least until he leaves the strictly scientific
handling, and strays into the fanciful or imaginative.

Wide knowledge, gained through language communication, ensures language acquisitions of the best kind—
those that are fully represented by meanings. The giving
and taking of names, by figurative transfer, is a purely
language manipulation ; but it proceeds, in the first instance, on the perceptions of things, and becomes a
matter of pure naming by synonyms, only when the
figures have been hackneyed by use.

To come back now to the position of the language
master. We have seen how he enlarges the pupils' vocabulary, by adding to their stock of equivalent names.
He might do this purposely and consciously, in every
ordinary reading lesson ; but the thing comes about, without express intention, in another way. I have already
remarked that for the early exercises in mere reading,
the reading-books provide easy, intelligible stories, or
descriptions, which give no strain to the attention, and
are not to be pressed as knowledge lessons. Now these
are calculated for lessons in language or naming. The

facts are familiar and easy; the language is choice and even adorned, being much above what the pupils are accustomed to in connection with the same subjects; and the expectation is that their language stores will be insensibly increased during the readings. To this end, the teacher contributes by the exercises that he founds upon the passages. In making the pupils remember the story, its connection and turns, he puts a series of questions to be answered nearly in terms of the book, or in varied terms, still of the better forms of diction, which they are supposed to be contracting in the course of their reading. Any teacher can make the most of this situation, if he will only realize to himself that the chief point of his teaching at this stage is language, and not the knowledge of things.

We are now at the point for considering the aids to language acquisition by committing passages to memory, more especially poetry. This may contribute to knowledge, but it is not the best way of imparting knowledge, and its value must be appraised rather as regards language. It is, however, one of the oldest devices of teaching; having the great merit of being plain and manageable, it is adapted to the lowest teaching capacity, and nobody can say that it is devoid of useful results. It certainly stamps upon the mind the material both of thought and of language, and they must be very hopeless subjects that cannot turn it to some account. In the schools of Greece, the children committed to memory, and recited, passages from the poets. Among the Jews the duty of teaching the children was, until a late period in Jewish history, imposed upon the parents, who could not adopt any higher method. Moreover the substance of the teaching was the Law, the Old Testament generally, and the traditions and rules of the Rabbis, none of this could be anything but memory work. In all modern schools, the practice exists more or less; in the French Lycées, at the present time, the culture of style is largely carried on through the practice of learning by heart the French classics. I say nothing at present of the learning of catechisms, hymns, and Bible passages, with a view to religious education.

Poetry has the natural preference in this exercise. The impressiveness of the measure, the elevation of the style, the awakening of emotion, favour its hold on the memory. Now a store of remembered poetry is a treasure in itself; its first effect is emotional, and its secondary uses are intellectual; it contains thoughts, images and language, of more or less worth, and such as are capable of taking part in our future intellectual constructions. Impassioned and rhythmical prose holds the next place; if it be inferior in form to poetry, it is yet more likely to be available in our own compositions.

To make poetry a source of pleasurable recollections in after life, it should be congenial from the first, and not too much of a mere task. The best of all our poetry stores are voluntary; to yield pleasurable emotion, an acquirement, like mercy, should not be strained. If it is to be made a task, it should be at an early age, when the despotic measures of the school are less taken to heart, and easily effaced. From seven to ten, the mind is in every respect more pliable to this particular work, than from ten to fifteen.

But now as to the intellectual worth of the acquirement, more especially for the end at present in view,

progress in the vocables of the language.   Granting the
facility in treasuring up compositions in poetry, we must
not be blind to its weaknesses.   The form, the compact-
ness, the feeling, the touches of lofty diction,—transport
us with the piece as a whole, without our troubling ourselves about the meanings of the parts, least of all, the
individual words.   It is only in the greatest masters,
that we are made alive to the sense of each word, and
not always in them.   Take as an example, the following
couplet, and note the sources of its impressions on a
youthful mind :—

> Thy spirit, Independence, let me share,
> Lord of the lion heart, and eagle eye.

It is quite enough to commend the couplet to a lusty
youthful soul, that it draws upon the egotistic feeling,
by the fact of sharing the spirit of some lord, no matter
who, or what, he is the lord of.   The meaning of ' Independence ' is not thought of at all ; nay, little is done
even to conceive the ' lion heart ' or ' eagle eye.'   The lines
would have their full inspiration, and would find a ready
admission to the young memory, if they were written—

> Thy spirit, Mumbo-Jumbo, let me share,
> Lord of the Tweedle-dum and Noodle-three.

It thus happens that poetry, above all other things,
may be committed to memory as three-fourths words,
and one-fourth meaning.   It is enough that a vague
thread of sense is traceable, provided interesting emotions are kindled in its track.

Prose passages are less easy to commit, but more
likely to be turned to account, than poetry.   It is not,
however, the highest economy to prescribe long compo-

sitions. What we want for ready use is a well-turned sentence form, or a suitable designation or phrase for some meaning that we are at loss to render. Now the stringing together of the sentences of a long passage does not contribute to the resuscitation that we desire; this is served more by the impressing of sentences individually, and especially such as have some marked and valuable characteristics, either in structure or in phrases. For very young pupils, these exemplary forms cannot be made apparent; and if they are to have their language memory strictly artificial, it can only be by the rote memory of passages. At the age of critical understanding, the committing of pieces at length should give place to the impressing of selected examplars, in the shape of single sentences or short series of sentences, made alive by critical exegesis or the singling-out of merits and defects. The pupils in the French Lycées should be looked upon as beyond the age when power of expression is best cultivated by the mechanical memory of passages however good. The practice with them savours of French drill and of inability to discriminate and criticize. The pupil will willingly absorb into his memory sentences and short passages that he has been awakened to appreciate and admire.

With advanced pupils, one of the best opportunities for committing to memory passages of poetry and exemplary prose, is in connection with exercises of recitation and delivery.

The foregoing remarks on the education in language by itself have had chiefly in view mere vocables, although the illustration has sometimes extended to the other part of language, namely, Structure. This part needs to be more closely viewed on its own account. In the practice of speech, in listening to speech, and in reading, we imbibe the structural arrangements of words in sentences and trains of sentences; and the passages that we learn by heart give us models of sentences as well as words and phrases. Long before we grammatically dissect a sentence, we are supposed to have been familiar with all the leading varieties of sentence forms.

Now the schoolmaster may allow the accumulation of sentence types to proceed silently with the reading lessons, or he may do something expressly to quicken the process of stamping them on the memory. I assume that the age of Grammar has not yet arrived; and hence the *science* of sentences is not entered upon. That age is, nevertheless, drawing near; and there may, conceivably, be a preparation for it, not to say a certain amount of independent tuition having the same final result, although not in the same complete form.

As has been said of vocables, so we may say of sentences, they follow the acquirement of meanings or thoughts. A fact needs a sentence to express it: a simple fact, a simple sentence; a complicated fact, a complicated sentence. 'The sun has set,' is a simple fact in simple sentence form. 'If you ascend to a height you will see the sun reappear,' is a conditional fact in a conditional sentence. If we have learnt, by verbal communication, many simple facts and many complicated facts, we have learnt many simple sentences and many complicated sentences. What more do we want? The answer is that, as with vocables, there is great convenience in knowing all the language forms for the same fact, simple or complicated. The learning of these additional, or supernumerary forms, is an education relating not to things, but to language.

As with vocables, sentence forms are best learnt in company with the knowledge that they are to express, and should not be made to precede that knowledge. The limitations to the principle have been sufficiently given.

Now as to the schoolmaster's province, in teaching or impressing these forms. The analogy of vocables still applies. The teacher having before him a given sentence, expressing a certain piece of information, can point out to his pupils, and exercise them in discerning and producing other sentence arrangements, with or without variation of the words employed. This is the best device yet promulgated for anticipating the formal teaching of Grammar ; only, it must be done upon system, although the system need not be obtruded on the pupils. When we come to put the question—What does Grammar (in our own language) do for us?—we shall find that this is one of its chief benefits.

Among the most simple examples of equivalence is the change from the Active to the Passive Voice—'Cæsar invaded Britain,' 'Britain was invaded by Cæsar.' Another is the interchange of Noun and Pronoun. We may further cite the conversion of Nouns into Noun clauses, of Adjectives into phrases and clauses, of Adverbs (single words) into phrases and clauses.

One of the most valuable preparatory exercises of equivalence is the filling up of omissions or ellipses, so common in every language as to be an authorized fact of the language. Half the difficulties of grammatical parsing grow out of these ellipses. 'Please to give me something to drink' is a grammatical puzzle till the full expression is given—'May it please you to give me something that I may drink.' The use of nouns as adjectives is altogether elliptical—'stone walls,' 'walls that are made of stone.' Another important contraction is the turning of clauses into abstract nouns—'What we see, we believe,' 'seeing is believing,' 'sight is belief.'

The *arrangement* of words and clauses in sentences admits of great variation. Qualifying words may either precede or follow the words qualified, but there is usually one arrangement that is best in the particular case. At the early stages. of the exercise, there is little attempt made to show preferences; the perception of the pupils is not sufficiently advanced; but opportunities should be taken of leading them on to this point, which is the goal of all language-teaching:

The teacher can form to himself a scheme of variations, for which Grammar and his own sense will be the guide. He will not iterate easy changes, nor harp upon such as are.devoid of importance. He will know what are the variations most needed in composition, and most adapted to bring out clearness and succinctness of expression; but he will not as yet divulge his motives or his reasons. Although it is enough for him to have in view the exigencies of Grammar, he may also ring a few of the rhetorical changes that are of common occurrence—as inversion of subject and predicate, interrogation, exclamation, metaphor and metonymy.[1]

---

[1] The methods of teaching by means of Equivalent Forms has been systematically and fully exemplified in a *First Work on English*, by Mr. A. F. Murison. The plan of the work is accommodated to a complete view of the Parts of Speech, and the Analysis of the Sentence; while it may be

By far the most searching equivalence of verbal forms is Obversion, or the stating of a fact from its other side; 'virtue is praiseworthy,' 'vice is blameworthy.' 'Thrice-armed is he that hath his quarrel just;' 'naked is he whose quarrel is unjust.' 'Heat favours vegetation,' 'cold retards it.' This passes beyond mere grammar; it is profoundly logical and also rhetorical; while as a discipline it is one of the most effective exercises that has hitherto been discovered. Quietly, unostentatiously, might a teacher now and then require this operation to be performed by a class in some favourable instance, and the result would be stimulating to a degree. It should only be done if the case carries with it its own explanation. 'If virtue is praiseworthy,' what are we to say of the opposite of virtue, vice? Any pupil of eight would be ready with a reply.[1]

TEACHING GRAMMAR.

The exact scope and suitable choice of these fore-shadowings of the regular Grammar will be rendered still more apparent by a full consideration of the greatly-vexed matter of Grammar teaching.

used either in advance of the grammatical definitions, or along with them. Especially valuable is the exemplification of Ellipsis, which contains the best key hitherto furnished to those numerous subtleties of our Grammar that originate in abbreviation.

[1] Take an easy example of the variations contemplated :—The sun is up, is risen, is above the horizon, has ascended, has mounted up the sky, has come into sight, has resumed his empire; is no longer down, beneath the horizon, concealed from view, dethroned. The verbal changes that may be rung upon the influence of the solar ray are still more numerous.

The synonyms for birth, life, and death are inexhaustible; they stretch away into wide regions both of fact and of imagination. From Homer downward, poets have been adding to the stock of expressive phrases for mortality.

We are too much given to supposing that the necessities and the benefits of Grammar are the same for our own and for foreign languages; yet the difference of situation is considerable. Before we begin our own Grammar, we have learnt, in a desultory fashion, the great body of what it teaches; when we begin the Latin Grammar, we find everything new. We could go on speaking and writing our own language very well without having ever seen a Grammar; we could not read a sentence in Latin, without some previous grammatical teaching.

This last condition might be evaded, by our being put through a course for a dead language similar to what we have gone through for our own; but the process is a clumsy attempt at reproducing a situation that cannot be reproduced; the only thing approaching to it, being the learning of a foreign living language by residence in the country. In beginning any new language, if we are at an age when the knowing and reasoning faculties are operative, our quickest course is to learn the Grammar. The reason is obvious. The Grammar abridges the labour, by generalizing everything that can be generalized.

No doubt the rules and the usages are sadly cumbered with exceptions, which make the acquisition burdensome; but that is not a sufficient reason for dispensing with them. Cobbett, in his self-willed way of doing things, proposed to set aside the rules for distinguishing the French genders, and to adopt a scheme of mastering them in detail by writing out all the nouns in the French dictionary. It could easily be shown not only that this would be a much greater strain upon the memory than the whole body of rules together with the exceptions, but that the learner would unconsciously resort to the plan of making rules for himself out of the uniformities that presented themselves in the operation; as, for ex- ample, the feminine gender of the abstract nouns.

Let us enquire closely into the real uses of Grammar; these will furnish the best guide as to the manner of teaching it.

1. The avoidance of the grosser forms of grammatical impropriety is a proper object of instruction, but does not need the amount of technical matter found in grammars of the present day. The vulgar errors in concord could be met by a much simpler means already adverted to. But the thorough expurgation of impro- prieties in the grammatical form of composition, could not be easily achieved without a much fuller scheme, in- volving what is the essential technicality of grammar— the laying out of the Parts of Speech, on which are founded Inflexion and Syntax. A few persons, ac- customed only to the best forms of 'the language, might approximate to a faultless style without grammar teach- ing; but not so the general mass. By the ear alone, we may be taught to avoid 'houses *is*'; but the insidious breaches of concord due to the distance of the subject and the verb—'the price put upon the houses *are*,' can hardly be explained without the terminology of Grammar. So, 'shall' and 'will,' and 'should' and 'would,' in commoner cases, may be correctly discriminated, but not in the more delicate or involved constructions, without grammatical assistance.

2. It is, to my mind, a circumstance of some value in favour of grammar that by it, for the first time, the

teacher isolates the attention of the pupils upon the language by itself. I have just been depicting a series of supposed lessons on language as such, before the age of grammar, and calculated to prepare for the formal teaching; but these lessons are of my own imagining; they have not yet been adopted in the current teaching. Teachers are feeling their way to a division of labour between knowledge lessons and language lessons prior to the formal teaching of the grammar; but speaking generally, there is as yet little attention paid to the structure of sentences, until forced on by the exercises in grammatical parsing. It is the grammar that solves the question—How shall we bring the pupils under a discipline in the forms of the language irrespective of the matter? Whether this is a good in itself, depends on the next position.

3. It is an aid to readiness, ease, correctness, and effectiveness of composition, to be led to examine the structure, arrangement, and constituents of the sentence. We may dispense with this training, but it will be to our loss; we shall not compass the arts of style so rapidly in any other way. This supposes a function of Grammar beyond avoiding censurable improprieties or violations of usage. That there is such a function has been already indicated. That consideration of equivalent sentence forms which might be induced without the aid of grammar, is almost compelled in the act of teaching grammar. Only a very mechanical system of parsing can keep clear of it.

This position ought to be self-evident. It may be illustrated thus. Suppose anyone at a loss to express a given meaning. The difficulty is due in the first instance to want of vocables, and next to want of resource in shaping sentences. How do we get experience on the last head? Partly, no doubt, by extensive reading, but equally by the habit of dissecting, putting together and varying the sentences that come up for review in language lessons. Teachers should be fully alive to the importance of the fact, and should conduct grammar parsing in accordance with it. That is to say, the words are to be parsed in their connections, and with reference to their functions, as principals or as qualifying words, and as to their efficiency for their ends, when compared with other equivalent words. This is to rehearse in advance the actual situation of the pupils when they shall have to engage in composition on their own account.

The Parts of Speech, and Syntax—in so far as it is concerned with the Analysis of Sentences, and Order of Words—are the portions of grammar most directly implicating sentence-structure.

4. Grammar contributes, in some of its departments, to the pupils' wealth in the vocables of the language. The iteration of examples everywhere has this effect; but the departments of Derivation and Inflexion are more expressly concerned. Under Derivation, considerable portions of the vocabulary pass under review, and are lodged in the memory. The comparison of Saxon with foreign elements is an exercise in naming. The study of prefixes and suffixes extends the familiarity with our means of expression, by showing the compass attained through the Composition of words. This has the recommendation of being free from all technical difficulties and abstruseness, for which reason it has been recommended as the branch best suited for a com-

mencement in grammatical training. Next, as regards
Inflexion. The lists of words brought forward in exemplifying gender and number make some impression, and
contribute to our readiness in remembering the words
when they are wanted. Teachers do well not to make
a point of burdening the routine memory with the
strings of words adduced in these subjects; there is
more likelihood of a lasting result being obtained by the
spontaneous and unavoidable dwelling upon them for
the exemplification of the rules, and this too is favourable to the recovery of the individual names for use
in composition.

Inflexion is the chief burden of the Latin Grammar;
there, it is the sole means of distinguishing the Parts of
Speech. Hence Latin Grammar is much easier—more
a work of memory and less a work of reason—than
English Grammar. Had Latin been our native tongue,
and English one of the dead languages, the proposal to
make the foreign Grammar precede our own would
have been denounced as monstrous.

The age for commencing Grammar.

Many persons are beginning to see the mistake of
commencing Grammar with children of eight or nine
years of age. Experience must have impressed teachers
with the futility of the attempt. Simplifications of
various kinds have been tried. Easy ways of presenting
the subject have been suggested to commence with; the
difficulties being postponed. Unfortunately for such
attempts, the difficulties lie at the threshold, and cannot
be evaded without rendering the entire subject a nullity.

24

The Parts of Speech cannot be understood at all unless they are understood fully.

When a pupil can be made to understand that a Sentence is made up of a Subject and a Predicate, that the predicate may be completed by an Object, and that subject, object and predicate may be qualified by secondary words,—a beginning may be made in Grammar. It is further to be desired that the logical notions of Individual, General and Abstract, should be understood by the pupil; for all these are essential to an intelligent grasp of the Noun and the Adjective. A certain amount of subtlety is needed to discern the meaning of words of *relation*, by which grammarians describe Pronouns, Prepositions and Conjunctions; and this subtlety supposes the pupil to have attained a certain age.

The suggestion is often made, and is probably acted on by some teachers, to teach grammar without book; on the assumption that the difficulties are not inherent in the subject, but come into being when it is reduced to form and put into the pupils' hands in print. There must be some fallacy here. What is printed is only what is proper to be said by word of mouth; and if the teacher can express himself more clearly than the best existing book, his words should be written down and take the place of the book. No matter what may be the peculiar felicity of the teacher's method, it may be given in print to be imitated by others, and so introduce a better class of books; the reform that proposes to do away with books entirely, thus ending in the preparation of another book.

Perhaps the teacher will reply that he does not propose anything absolutely original, but merely to select such points as the pupils can understand, being guided by his natural tact and judgment as to what he finds to succeed. Even so, it is quite possible to embody this selection in a permanent form; and what is good for one class is likely to answer for other classes at the same stage. Again, it may be said that the children are not of an age to imbibe the doctrines from a printed book, but can understand them when conveyed with the living voice. There is much truth in this, but it does not go the length of superseding the book, which will still have a value as a means of recalling what the teacher has said, and as the basis of preparation to answer questions thereon. If a class is to be taught purely *vivâ voce*, its progress must needs be very slow; the proceeding belongs to the infantile stages, when slowness is not an objection.

To teach Grammar without a printed text, is like teaching Religion without a manual or catechism; either the teacher still uses the catechism, without the print, or he makes a catechism for himself. There can be no teaching except on a definite plan and sequence, and good, instead of harm, arises from putting the plan in print. The grammar teacher, working without books, either tacitly uses some actual grammar, or else works upon a crude, untested, irresponsible grammar of his own shaping.

Taking English Grammar as a whole, easy parts and difficult together, I venture to think that it cannot be effectively taught to the mass before ten years of age. To smooth over the asperities, and to pick out what happens to be simple, in order to adapt it to an earlier age, is not to teach the subject in its proper character, but as a mongrel compound, half-understood and quite inade· quate for the ends of grammar. It is the worst economy to anticipate the mind's natural aptitude for any subject; and the aptitude for Grammar in its true sense does not exist at eight or nine years of age. I have already ex· pressed the opinion that it is more difficult than Arith· metic, and is probably on a par with the beginnings of Algebra and Geometry. Commenced at a ripe age, not only is the tedium of the acquisition vastly reduced, but the advantages are realized in a way that is impossible when it is entered on too soon.

This postponement is open to one and only one real objection, so far as I am aware. It leaves a gap in the teaching that there is some difficulty in filling up. If the teacher is to exclude grammar, he must exclude English exercises entirely, and make the whole of his teaching, as far as concerns the reading part, consist of knowledge lessons, in which region also he often incurs the evil of attempting matters too high for the pupils at the time. There would seem to be an absolute necessity for contriv· ing lessons in English, whether amounting to grammar or not. The difficulties of grammar are the difficulties of all science—generalities couched in technical language; and there is a possible preparation of the concrete and the empirical here as in other sciences.

This brings us back to what has been said already as to the province of the teacher in regard to expres· sion as distinct from the thought or meaning; namely, exercising the pupils in equivalent forms, while at the same time adding to their stock of vocables by practice in synonyms. Whether or not this be a direct prepara· tion for grammar, it is a preparation for the end of grammar—the power of composition—and would not be lost even if the regular or technical grammar were never reached.

The more special preparation for the formal study of the grammar would be, for one thing, to exemplify the division of a sentence into its subject and predicate, without the use of these formidable words. *Apropos* of a sentence of information about something—'the fox is a very crafty animal '—it is easy to ask what the saying is about? The fox. What is said about the fox? It is a very crafty animal. To this might be added exercises in the names of objects, so as to bring out the difference between individuals and classes, and to show how the class noun comes, and how it is narrowed by an adjective into a smaller class. These logical distinctions might be started on the eve of entering grammar, say a few months in advance. They are a small contribution to genuine logic, and it is only through them that grammar can be cited as the means of a logical training. Provided with such a discipline, the pupil can make an effective beginning with the Parts of Speech, by grappling with the Noun, its definition and its kinds ; while the other logical notions that lie ahead may be left till they come up. I will instance further the important distinction between co-ordination and subordination, without which the relative pronouns and the conjunctions are dark where they ought to be in a blaze of light.

Two years before the age of grammar, the lesson in English might be isolated in the reading. This is the only way to give it a clear *locus standi* in education. Putting a grammar question or two during an informa-

tion lesson is a kind of trifling. If the information is of
any consequence, it needs the attention to be kept well
upon itself; if the language lesson is in earnest, it equally
wants concentration of mind; and the rapid shifting to
and fro rapidly between two totally different studies is
adverse to both. A reading lesson is (1) a lesson for the
mechanical art of reading, together with spelling, (2) a
lesson in information, to be understood and remembered,
and (3) a lesson in language. For a long time, the first
lesson is the only. thing considered : by and by, the
second—the information—is taken in hand, and becomes
a more and more engrossing part of the teacher's work,
carrying, as a necessity, language with it. The third
stage—language by itself—is the latest of the three; and
needs a special handling, which will not be given, unless
it have a certain hour allotted to it alone. The information passages may be used for the language lesson, but
the information is not to be adverted to further than is
necessary for considering the language ; while there
may be passages little suited for information, and well
suited for language; such are extracts—poetry and prose,
belonging more to the *belles lettres.* The lessons so isolated would be driven to take shape, and continuity ;
they would inevitably form a course, each lesson following on its predecessor. The teacher would have to make
up his mind to a plan, and would not be at a loss to find
some such substitutes for the beggarly grammatical elements as I have endeavoured to point out. The occasional committing of short elegant passages to memory
might be associated with the language lesson.

When the age of Grammar is reached, the problem

of teaching it solves itself. It is a practical science, having general principles which become rules ; these need to be explained and applied to the particular cases. Instead of adopting devious routes to escape difficulties, the teacher now follows the direct course as chalked out by the concurrence of the best grammarians. There is still considerable variation of views as to the working out of the details ; and a few remarks may here be offered on the more important discrepancies.

1. I hold that the subject of Inflexion should be separated from the Parts of Speech. The defining, classing, and exemplifying of the Noun, Pronoun, &c. constitute one distinct and homogeneous operation ; the inflecting of the inflected parts is quite a different subject, and is best prosecuted consecutively and without interruption.

2. The ' Analysis of Sentences,' which has been the turning-point of the radical reform in the Definition of the Parts of Speech, is not yet pushed to its legitimate conclusion in amending our Syntax. It enables us to take a sentence to pieces, and it puts qualifying phrases and clauses in their true light as equivalents of nouns, adjectives, and adverbs; but it leaves out of view the consideration of the right ordering of the sentence through the proper disposition of the qualifying adjuncts. Yet this has more to do with good composition, than all the rest of the grammar put together.

3. There is great interest, and some utility, in tracing the course of our language from the more ancient dialects, but this subject may easily run to a disproportionate length in the first stages of English teaching. Present meaning and use are the only guidance to the employ-

ment of the language ; the reference to archaic forms can sometimes account for a usage, but cannot control it.

THE HIGHER COMPOSITION.

Grammar and Rhetoric, or the Higher Composition, are not separated by any hard and fast line; yet the two departments are distinct. To be grammatical is one thing ; to be perspicuous, terse, or unctuous, is another thing.

In the view we have taken of grammar teaching, results far beyond mere correctness are attained. Nevertheless there is still a large domain of instruction in Style; on entering which new methods are called into play.

Rhetoric, like grammar, has its rules, which are to be understood, exemplified, and carried into practice in composition. Moreover, these rules must be embodied in a systematic array; which supposes numerous explanations and definitions of important terms. The whole subject is divided into two Parts—one on Style in General, or the explanations, rules, and principles, applicable to every kind of composition ; the other on the special Forms or Kinds of Composition, as Description, Narration, Exposition, Persuasion, Poetry.

In a separate work ('English Composition and Rhetoric'), I have indicated what I consider a suitable arrangement of the details of the subject, and have also brought together and exemplified all the maxims and rules that I consider valuable. In commencing the first part (Style in general) with the Figures of Speech, I am guided by the universal recognition of certain leading designations, under which many of the Rhetorical principles are brought forward in advance. In point of fact, the Figures, well explained, are of themselves a short course of Rhetoric. The other matters required in a complete view of the Laws of Composition in general, are—the Qualities of Style, and the laws of the Sentence and the Paragraph.

The explanation and exemplification of the various terms employed, and of the rules and principles of composition, would seem to indicate with sufficient clearness the course to be pursued in the higher department of composition. Still, there is a certain latitude in the choosing of exercises, and the practice of teachers is very various in that respect. We may, therefore, offer a few words on the point.

In the case of young pupils, there are very strong objections to Essay or Theme writing It is a contravention of the all-pervading principle of teaching—to do one thing at a time. The finding of the matter absorbs half or more than half the attention of the learner, and leaves little room for the study of the style. Besides which, the writer necessarily travels over a wide compass of expression, and it is impossible for the master to take notice of all the faults and inadvertences ; while it is scarcely practicable to conduct a class, or impart simultaneous criticism, by means of essays.

In a composition exercise, the matter should be provided, and the pupils required to find a suitable expression. Something might be given in outline, which they are to expand ; but even this is too much of a subject lesson at the early stages. The conversion of poetry into prose is a very convenient exercise ; the danger is that, in stripping off the poetical form, the pupil does not leave enough of energy and elegance to make good prose. A still better exercise, although less ready to hand, is to change the form of a given prose passage on some definite plan, arising out of the rhetorical lessons going on at the time:—to remove or insert figurative terms of expression ; to pare down redundancies, or to supply in a too curt passage some needful expansion ; to re-arrange sentences, upon definite principles ; to alter the proportions of the Classical and the Saxon words ; to vary, in all the best ways, the modes of expressing the same thing.

The standing devices connected with Agreement and Contrast should be well iterated. Under Agreement, come Example and Similitude ; Contrast is the universal remedy of vagueness, and one of the chief arts of giving point to language.

How to order sentences in a paragraph is a high and arduous undertaking. It is best studied upon the passages that occur in reading ; which passages may be prescribed for rearrangement, according to principles laid down. The making of a good paragraph is nearly the highest feat of orderly expression as such. The arrangement of a discourse contains scarcely any new difficulties of mere composition.

The exercise that seems to me to comply best with the requirements of the composition lesson, is the critical exegesis of good prose and poetry passages, conducted along with a course of rhetorical instruction. I have given abundant examples of this in another place (' English Composition and Rhetoric '). The pupil's mind, in these lessons, is wholly bent upon the ways and means of expression ; and I scarcely know any other exercise

that is equally recommendable on the same vital circum·
stance.

The whole gist of rhetorical teaching, as thus viewed,
is to awaken the minds of the pupils to the sense of
good and evil in composition. This I take to be the
prime requisite. For, although in order to write well, a
command of expression is even more necessary than the
power to judge of good writing ; yet, the teacher can do
but little for the one, and can do a great deal for the
other. Affluence of language is the fruit of years ; very
many of the niceties and delicacies of composition may
be made apparent in a six months' course. On the supposition, however, that a portion of time is continuously
devoted to the English Language for a series of years, a
vast deal may be done to impart both abundance of
phraseology and the effective employment of it.

There is a preparation for the formal and methodical
teaching of Rhetoric and Style, analogous to the preparation for grammar ; namely, to vary rhetorically the
occurring modes of expression, and to indicate the better
and the worse, without reasons. In some points, the
pupils' own feeling of the superiority of one form as
compared with another, would come into play, and
might receive direction from the master. The variations
caused by the presence and the absence of Figure, the
changes in the arrangement of Sentences, would be felt
before the age of rhetorical system.

In the course of promiscuous reading, the pupils
might be gradually awakened to such leading Qualities
of Style as Clearness, Strength, Pathos. By well-
selected instances, they might be made to discern the
difference between Simplicity and its opposite, also be-

tween Strength and Feeling or Pathos. A farther step would consist in calling attention to the methods and arts of attaining those effects ; although, apart from regular rhetorical instruction, this could not be carried out with advantage.

ENGLISH LITERATURE.

The teaching of English Literature is liable to all the perplexities attendant on the framing of a course of History. It is a mixture of what is easy, intelligible, and interesting to the youngest, with what is technical and abstruse, and accessible only to the mature mind. Just as in General History, there is no possibility of contriving a course that shall in every point keep the steady level of the juvenile capacity.

Our great authors can be arranged in an interesting Chronology, and this might be fixed in the memory, long before the characteristics of each could be understood. Their lives also could be read, as narrative interest ; including also the mention of their works, the dates and subjects of these, with a few necessarily vague expressions respecting their merits. This is scarcely lesson work ; it is rather the amusement of growing minds.

The History of Literature, narrowed to its strict domain, is the criticism of literary works in all that relates to style or composition. What makes the History is the regarding of our English authors in a connected series, each having more or less relation to the preceding. This historical treatment of Literature is itself a branch of the Belles Lettres, being always conducted with studious regard to the graces of composition.

The basis of literary criticism, whether of detached authors, or of the literary succession, is rhetorical knowledge, or an exact acquaintance with the qualities and the laws of style, gained in the manner that we have above sketched. The great fault in the early teaching of English Literature, is to address it to minds so little acquainted with literary qualities as not to comprehend the meaning of the terms employed. After the rhetorical nomenclature is properly unfolded, criticism and history are self-explaining.

At the present time, the teaching of Literature in schools has taken the form of the study of selected works of the greatest English authors, from Chaucer downwards. There is now provided an ample series of such works, with every needful aid in the way of commentary or annotation. Two points have to be considered in connection with the working of this method: the first as to the selection of authors, the second as to the way of handling them.

I. In the selection, the later authors are to be preferred to the earlier, and the prose authors to the poets. The first of these two maxims proceeds on the fact that English prose style has improved, and is improving ; while the thoughts and the general interest are still more in favour of the moderns. Hooker and Bacon and Temple, were in their day great writers of prose ; but, for our purpose, they are surpassed by Burke and Hall and Macaulay. The pupil, at the outset, should see prose at its very best ; and should be led backwards to the less perfect examples. The interest of many of the older prose writers, although not entirely exhausted, undergoes an almost steady decrease with the lapse of time ; and only in sample are they fit to occupy the hours of an English class. It is in setting forth the History of Literature, considered as the development of the literary form, that they have their most suitable place.

The value of style does not depend on the matter conveyed, but the interest of the subject has much to do with the impression made by the language. If the thoughts have become effete, if the subject, whatever it is, has been much better handled by later authors, our attention flags, and none but extraordinary merits of style can detain us. Moreover, style comes home to us with most effect, when it is accompanied by matter that we are ready to cling to for its own sake.

The second maxim—to give the preference to prose authors, in early teaching—proceeds on the practical consideration that prose is what we habitually employ; while poetry is for our enjoyment, like music and paint-ing. If it is not waste of time, it is at least great dis-proportion, to keep a class occupied for months on a play of Shakespeare, or on three Books of 'Paradise Lost.' No doubt many of the exercises performed on a prose lesson can be performed on poetry; and moreover, the greatest efforts of style as such are put forth by the poets. Yet the argument is unanswerable, that if these exercises are to improve our own composition, it is as prose composers ; and for a good model of prose we must refer, not to a poet, but to a writer of prose.

Unless the space allowed for English is very con-siderable, as it might be if Classics were displaced from the higher education, poetry can come in only by selected passages.   It must be referred to in Rhetorical teaching, as exemplifying the Qualities and the Arts of Style; and that is as much as, in my judgment, should be attempted.   We may admire Chaucer, Shakespeare, Milton, and Pope, but they are not the one thing needful in an English class.   The best English teaching would say little about them at the time, but would, nevertheless, give the pupils the aptitude and the zest for reading them when they have left school.   Not one of these writers is child's play.   None of them can be read with any tolerable appreciation before eighteen or twenty; aud the full enjoyment of them is much later.

II. As regards the best mode of using the selected works of our great authors, I have to fall back once more upon the great principle of Division of Labour— the separation of the language from the matter.   A portion of Bacon, of Addison, of Burke, of Macaulay, —may be a knowledge lesson, or it may be a language lesson.   In present practice, it is apt to be both.   But, as I have said, the English teacher should have nothing to do with the matter, except in relation to the manner. He may read with his pupils Burke on the French Revolution, but .he should not trouble them with the political thoughts, but only with the conduct and method of the exposition—with the sentences, the paragraphs, the illustrations, the figures, the qualities, the diction.   He does not need to make them con the entire treatise, with its interminable repetitions.   It is his business to indicate important peculiarities in the expression and in the handling— what to imitate and what to avoid in the one or in the other.   When he has got out everything of this kind that the work can yield, he

has done enough. It is not his business to teach poli‧
tical philosophy; and if it were, a much better handbook
could be found for beginners in that subject.

The teacher of advanced classes in English does
not even undertake to explain difficulties or obscurities
of meaning, except to point a language lesson. It is
doubtful how far he should take upon himself to explain
figurative allusions; he certainly should not charge him‧
self with interpreting the far-fetched comparisons of
florid writers and poets, nor make these the occasion for
giving desultory information in history, mythology, geo‧
graphy, natural history, manners and customs. Such
explanations are suitable in those early reading lessons
wherein meaning and language are not yet differentiated.
But in the later stages of instruction in style, such things
are to be forborne. General information is now given
in most subjects by systematic teaching; and the mis‧
cellaneous contributions from the allusions of poets are
superseded by a more excellent way. Pupils need not
follow out the references to the similes of Milton farther
than to feel their force; and such as need much explain‧
ing may be passed over. The pressing matter is, to be
led to discriminate the effects of the composition, and
to see what are the arts that bring about these effects.

The same rigid principle of division of labour would
exclude from English teaching whatever relates to the
history, manners and customs of the country, and all
occasions for calling forth patriotic and moral sentiment.
Such matters obviously belong to historical and other
teaching, and should not be encroached upon by the
English master any more than by the Drawing master.

CHAPTER X.

THE VALUE OF THE CLASSICS.

THE chapter on Education Values was purposely left incomplete; the vexed question of the study of the Classics demanding a separate and full discussion. As respects the Higher Education this is the most important of all the questions that can be raised at the present time. The thorough-going advocates of Classics hold Latin and Greek to be indispensable to a liberal education. They do not allow of an alternative road to our University Degrees. They will not admit that the lapse of three centuries, with their numerous revolutions, and their vast developments of new knowledge, make any difference whatever to the education value of a knowledge of the Greek and Roman classics. They get over the undeniable fact, that we no longer employ these languages, as languages, by bringing forward a number of uses that never occurred to Erasmus, Casaubon or Milton.

In the Middle Ages, the use of Latin was universal. After the taking of Constantinople, Greek literature burst upon Western Europe, and so entranced the choicer spirits as to bring about a temporary revival of Paganism. To the Christian scholarly enquirer, Greek was welcomed as laying open the original of the New Testament, to-

25

gether with the Eastern Fathers of the Church. The zeal thus springing up rendered possible the imposition of a new language upon educated youth, which might have well seemed too much for human indolence. Our Universities accepted the addition; and the teachers and pupils had to speak Latin, and read Greek.[1]

The men of the fifteenth and sixteenth centuries had their own follies, errors, and superstitions; but their mode of estimating the worth of the classical tongues was plain common sense. Says Hegius, the Dutch scholar (master of Erasmus, head of the College of Deventer, 1438–1468): 'If anyone wishes to understand grammar, rhetoric, mathematics, history, or Holy Scripture, let him read Greek. We owe everything to the Greeks.' Luther advocated the new learning, in his own vehement way: 'True though it be that the Gospel came and comes alone by the Holy Spirit, yet it came by means of the tongues, and thereby grew, and thereby must be preserved.' Melancthon regarded the languages solely as means to ends, and his scheme of education embraced all the departments of knowledge on their own account. Hieronymus Wolf, of Augsburg, was emphatic on the same point: 'Happy were the Latins,' he says, 'who needed only to learn Greek, and that not by school-teaching, but by intercourse with

[1] 'Thus in the Middle Ages Latin was made the groundwork of education; not for the beauty of its classical literature, nor because the study of a dead language was the best mental gymnastic, or the only means of acquiring a masterly freedom in the use of living tongues, but because it was the language of educated men throughout Western Europe, employed for public business, literature, philosophy, and science, above all, in God's providence, essential to the unity, and therefore enforced by the authority, of the Western Church.'—(Mr. C. S. Parker, in *Farrar's Essays on a Liberal Education*, p. 7.)

living Greeks. Happier still were the Greeks, who, so soon as they could read and write their mother tongue, might pass at once to the liberal arts and the pursuit of wisdom. For us, who must spend many years in learning foreign languages, the entrance into the gates of Philosophy is much more difficult. For, to understand Latin and Greek is not learning itself, but the entrance-hall and antechamber of learning.' (Parker.)

That the value of a knowledge of the classics, on the ground of the information exclusively contained in Greek and Latin authors, should decrease steadily, was a necessary result of the independent research of the last three hundred years. The rate of decrease has been accelerated during the last century by the abundance of good translations from the classics. In this progressive decrease a point must be reached when the cost of acquiring the languages would be set against the residuum of valuable information still locked up in them, and when the balance would turn against their acquisition. In the meantime, however, other advantages have been put forward that are considered sufficient to make up for the loss of value brought about by the causes now mentioned.

I. *The Information still locked up in Greek and Latin Authors.*

This is the professional argument, but the case respecting it is so very obvious that we can hardly be too brief in presenting the matter.

That there is not a fact or principle in the whole compass of physical science, or in the arts and practice

of life, that is not fully expressed in every civilized modern language, will be universally allowed. There will
not be quite the same consent as regards moral and
metaphysical science ; it being contended that in Plato
and in Aristotle, for example, there are treasures of
thought that never can be separated from their original
setting in the Greek language. Again, the ancient
literatures are the exclusive depositories of the historical and social facts of the ancient world ; but all this
is eminently translatable, and has been abundantly reproduced in the modern tongues. A certain exception,
however, is made here also, namely, that for the inner
or subjective life of the Greeks and Romans, the best
translations must still be at fault.

As regards Greek philosophy, it may be safely said
that its doctrinal positions and subtle distinctions are at
this moment better understood through translators and
commentators, writing in English, French, and German,
than they could have been to Bentley, Porson, or Parr.
The truth is that, in translating, a knowledge of the
subject is at least co-essential with a knowledge of the
language. When the Professor of Greek Literature, in
Cosmo's Platonic Academy at Florence, lectured on
Plato, the Latin Aristotelians asked with indignation
how a philosopher could be expounded by one who was
none himself.

That the inner life of the Greeks and Romans cannot be fully comprehended unless we know their own
language, is a position that gives way under a close
assault. The inner life must be understood from the
outer life, and that can be represented in any language.
Whatever sets well before us the usages, the modes of

acting and thinking, the institutions, and the historical incidents of any people, will enable us to comprehend their inner life, as well as can be done in surveying them at a distance ; and all this is quite possible through the medium of translators and commentators.

This seems enough as far as concerns the professions. In medicine, for example, it will not be contended that there is anything to be gained by classical scholarship. Hippocrates has been translated. Whatever Galen knew is known independently of his pages. But indeed, only a purely historical value can attach to any medical work of the ancient world.

Again, the lawyer can obviously dispense with Greek. There may be a certain claim made for Latin in his case, in consequence of our position with reference to Roman Jurisprudence. But this too has been sufficiently represented in English works to make the whole subject accessible to an English reader. The Latin terms that have to be retained as untranslatable by single words in English can be explained as they occur, without anyone requiring to master the entire Latin language. As to the power of reading Latin title-deeds, if one man in a business establishment possesses it, that is enough.[1]

The plea for classics to the clergy has always been accounted self-evident and irresistible. Even here, however, there are qualifying circumstances. It is the business of a clergyman to understand the Bible, which involves Hebrew and Hellenistic Greek. Classical Greek and classical Greek authors are not necessary ;

_____

[1] Mr. Sidgwick says a lawyer 'ought to be acquainted with Latin grammar, and a certain portion of the Latin vocabulary.' The necessity for the grammar is not self-evident.

while the utility of Latin extends only to the Latin
Fathers, the scholastic theology, and the learned theo-
logians of the Reformation, including Luther, Melancthon, Calvin, and Turretin.

Now there is no book that has been so abundantly
commented on as the Bible. Every light that scholarship can strike out has been made to shine through the
vernacular tongues ; there is scarcely a text but can be
understood by an English reader as the ablest scholars
understand it ; and the study of the original languages
must be prosecuted to a pitch of first-rate scholarship
before anything can be gained in addition to what everyone may know without scholarship.

Among the caprices of opinion on the present question may be ranked the very slight stress that is put
upon the Hebrew language in the education of the
clergy. The most exacting churches receive a candidate for orders on a very easy Hebrew pass; and it is
never supposed that more than a small number of
preachers in any church habitually consult the Hebrew
Bible. Yet the Old Testament, containing as it does a
large mass of sentiment and poetry, and referring to a
state of society far removed from our own, is one of
the books most difficult to exhibit in translation. Granted
that, as respects the Old Testament, there may be an
unexhausted, possibly an inexhaustible, suggestiveness
in the knowledge of the original tongue, the fact remains
that inattention to Hebrew is all but universal ; while, as
respects the New Testament, a knowledge of the original can scarcely add anything to the ample exegesis
provided by theological scholars. Whitfield knew no
Hebrew and little Greek.

The Hellenistic Greek of the New Testament does not involve classical Greek authors. It might be taught like Hebrew in the Divinity schools, and entirely disconnected from the literature of Pagan Greece. That these Pagan authors should be nursing fathers and nursing mothers to the Christian Church, is a standing wonder. That Christian youth, so carefully withheld from the language of sexual impurity, should be allowed such a liberal crop of wild oats as a course of classical reading supplies, is not less wonderful.

The natural course as regards the clergy would be to encourage a small number of scholars to prosecute the study of the original languages of the Bible and all the allied learning, and to dispense with these languages as regards the mass of working clergy, who may turn their time to more profitable account.

II. *The Art Treasures of Greek and Roman Literature are inaccessible except through the languages.*

It must ever remain true that certain artistic effects of literary composition, and more especially poetry, are bound up with the language of the writer, and cannot be imparted through another language. These very peculiar effects, however, are not the greatest in themselves, nor the most valuable for literary culture. The translatable peculiarities far transcend in value the untranslatable; if it were not so, where should we be with our Bible? Melody is the most intractable quality; of this alone can little or no idea be imparted by translations. Even the delicate associations with words can be expounded through our own language; just as they must be to the pupil who is studying the original. As regards all dead languages, much of this subtle essence must have vanished beyond recovery. Learning Greek does not put one in the same position to Homer and Sophocles, that learning German does to Goethe. All that a scholar can know he may find means of imparting to one that is not a scholar.

The subtle incommunicable aroma of classical poetry is one of the luxuries of scholarship. The mass of students cannot reach it; and it may be bought too dear. Moreover, the translatable virtue of the great poets is so great, that we may have many a rich feast, through translations alone: witness the enthusiasm for Pope's 'Homer.' Horace is perhaps the most untranslatable poet of antiquity; but the difficulty has been a stimulus to marvels of verbal dexterity in approaching the original; and he that is conversant with the translations now accessible to the English reader, cannot be far from the kingdom of heaven.

III. *The Classical Languages train the mind as nothing else does.*

This argument was not advanced in the days when the dead languages were useful in their character as languages; either it was not felt in the sixteenth and seventeenth centuries, or it was unnecessary. That it is so much relied upon now, is tantamount to a surrender of the previous arguments, or at least suggests doubts as to their sufficiency. It has that amount of vagueness about it that would make a convenient shelter to a bad case We must ask specifically what the training consists in.

For one thing, there is abundant employment given to the memory; but the proper word for this is not 'trained' but 'expended.' A certain amount of the plastic force of the system is used up, and is therefore not available for other purposes. This is the cost of the operation, for which we have to show an equivalent in solid advantages.

The faculties supposed to be trained are the higher faculties named Reason, Judgment, and Constructive or Inventive Power; and the exercises reckoned upon to give the training are conning grammar, and translating.

The influence of Grammar can soon be told. To learn Grammar is, besides employing memory, to understand certain rules and to apply them as the cases arise, bearing in mind the exceptions when there are any. Inflexion is the easiest part. Latin nouns in *a* of the first declension are declined according to a type; one example is given, as *penna*, and the pupil has to adhere to the type with *femina* and the rest. This represents the operation that is requisite whenever we can rise from particulars to general knowledge. 'A fine day,' 'a good road,' 'a boiling kettle,' 'a loaf of bread,' are general ideas that are connected with practical injunctions, and whoever has to comply with these injuctions must understand the ideas and apply them as the occasion serves. Sometimes the notion is accessible to the weakest capacity, sometimes it is the reverse; there are all degrees of difficulty up to the subtleties of professional lore, and the abstruseness of science or philosophy. The chief point is, that no branch can have a monopoly of the exercise of seeing the general in the particular; we cannot evade the necessity of the task. Whether one sub-

ject is better than another for our education in the matter
depends upon whether it is possible to ease the labour
of conceiving the more difficult abstractions by something foreign to them; whether mathematics or metaphysics can be made easier by toiling in some foreign
lines of thought, as Latin Grammar, English Grammar,
or Botany. It remains for anyone to show that such an
influence exists; the arguments for the efficacy of grammatical discipline do not reach the point; they assume
that grammar has a monopoly of exercising the mind
upon generalities, a point that has yet to be proved.

Grammar as exemplified in the Latin and Greek
languages is particularly devoid of subtlety, until we
come to certain delicacies of syntax, as in the construction of the tenses and moods of the Verb. The Parts
of Speech are assumed without any definition ; they are
recognized by the Inflexion test, and not by their function in the sentence ; being in that respect very different
from what is found in English Grammar. This has been
made an argument for taking Latin before English—the
easy grammar before the abstruse one. But the greater
should imply the less. If, at the proper age, a pupil has
mastered English Grammar, he has, in point of reasoning
power, gone a step beyond Latin or Greek grammar, and
should therefore be relieved from further labour for perfecting his reasoning faculties in the grammatical field.

It is in the exercise of translating from Latin or
Greek into English, and *vice versâ*, that the highest
mental efforts are made, and the greatest strain put
upon the faculties. Accordingly, it is to this exercise
that the supposed training more especially applies. Now
the mere conquering of difficulty is not special to any

line of study; we must further enquire what are the special difficulties to be overcome. The exercise of translating is a constructive effort: given a passage, a certain amount of grammatical and verbal knowledge, and the use of a dictionary, the pupil has to divine the meaning. There are three stages in the pupil's progress. In the first, his information and resources are unequal to the task, in which case the labour can do him very little good; we are not the better for working at a point where we cannot make any progress. The second stage is where, by a certain measure of application, the pupil can succeed; in which case, the operation is exhilarating and rewarding, and will be achieved. The highest stage is when the work can be performed with ease, and without any effort at all; in which stage there is no difficulty to be overcome, and, therefore, very little effect accruing from the exercise. We are to assume, what is not always the case, that the student can be uniformly placed in the second situation, and are to enquire what there is in the particular work to train, discipline, or strengthen any of the higher faculties.

The translation exercise is a tentative process; the meanings of the separate words have to be ascertained ; and out of several meanings of any one word, a selection has to be made such as to give sense along with the selected meanings of the others. Various combinations have to be tried; baffled at one attempt, the student must make a second and a third, until at last he alights upon something that pays a due regard to every word and every peculiarity of grammar. A considerable amount of patient effort is demanded, and the long-continued exercise of patient effort must do something to

form habits of application. There is not, however, anything specific, unique, or unparalleled in the operation.
All study whatsoever needs a similar exercise of patient
application; and many kinds of study take precisely the
same form, namely, assigning to words alternative meanings, until some one meaning is hit upon that resolves a
difficulty. It is the application needed to solve riddles
and conundrums. To make out the meaning of a scientific proposition, to find the rule that fits a given case,
we must try and try again; we reject one supposition
after another as not consistent with some of the conditions of the problem, and remain in patient thought
until others come to mind.

It is in the interpretation of language that most
difficulty is felt in keeping the pupil always in the medium position above described; giving him work to do
that shall neither exceed his powers, nor be too easy to
call them into full exercise. With a passage that the
dictionary does not give the means of rendering, the
chance is that the attempt will not be seriously made, so
that the mind is not put on the *qui vive* to drink in with
avidity the master's explanation. It is, moreover, generally admitted that the use of 'cribs' does away with
the good of the situation, as regards translating into
English. Hence to secure any discipline at all, the
operation of translating from English into Latin and
Greek must be kept up, although in itself the least
useful of any.

The remark could not fail to be made that the operation of translating is necessarily the same for ancient and
for modern languages; and, therefore, any modern language yields whatever discipline belongs to the situation.

It cannot avail much, in reply, to advert to the peculiarities of the Latin and Greek Grammars—the more highly inflexional character of the languages ; for each language has its specialities, and the business of the pupil simply is to attend to them. Every language must express the same facts of time and manner, and it cannot be very material, as far as regards mental discipline, whether it is by inflexion or by auxiliaries. The fact of inflexion is sufficiently experienced in any case ; and how far it is carried is an inferior consideration.

In Science, far more than in Languages, is it possible to adjust the difficulties at each stage to the strength of the pupils, although, undoubtedly, to do this in any subject needs very good teaching. The Grammar of language being most nearly allied to science, can be best graduated in this way ; while, in the miscellaneous chances of translation, difficulties start up without any reference to order or the preparation of mind of the pupils, and the thing cannot be otherwise.

The argument from Training is applied to certain special points, some of which will be considered under separate heads: such are the discipline in English and in Philology generally. Much stress is laid upon the remark that it is necessary to know more languages than our own to be delivered from certain snares of language; and the favourite example is the ambiguity of the verb ' to be.' It so happens, however, that this very ambiguity—predication and existence—was pointed out by Aristotle (*Grote's Aristotle*, i. 181).[1]

---

[1] In an address to the Social Science Association in 1870, Lord Neaves recommended the study of Latin, Greek, and French, as the best means of cultivating precision of thinking. Now, whether or not the writers in

In the interesting Rectorial Address of Professor
Helmholtz, delivered this year to the University of Ber-
lin, the merits and demerits of the different academical
institutions of Europe are freely indicated. With reference to the English Universities, Oxford and Cambridge,
the professor thinks his own countrymen should endeavour to rival them in two things. 'In the first place,
they develop in a very high degree among their students,
at the same time a lively sense of the beauties and the
youthful freshness of antiquity, and a taste for precision
and elegance of language; this is seen in the fashion in
which the students manage their mother tongue.' This
must refer to the prominence still given to the classics in
Oxford and Cambridge; yet, in Germany, the classics are
far more studied than in England, whether we consider
the universal compulsion of the Gymnasia, or the special
devotion manifested by a select number at the Universities. Whatever good mere classical study can effect
must have reached its climax in Germany. As regards
Oxford and Cambridge, and particularly Oxford, the
best parts of the teaching seem to be those that depart
most from the classical teaching, as, for example, the
very great stress laid upon writing a good English essay.
It is often said, that even in a professedly classical
examination, a candidate's success is more due to his
English Essay than to his acquaintance with Greek and
Roman authors.

After refuting a number of the alleged utilities of
classical learning, Mr. Sidgwick still reserves certain dis-

those languages are distinguished above all others for precision, it is a singular fact, that these are the languages of the three peoples most remarkable for confining their attention to their own language.

tinct advantages as belonging to the study of language. 'In the first place, the materials here supplied to the student are ready to hand in inexhaustible abundance and diversity. Any page of any ancient author forms for the young student a string of problems sufficiently complex and diverse to exercise his memory and judgment in a great variety of ways. Again, from the exclusion of the distractions of the external senses, from the simplicity and definiteness of the classification which the student has to apply, from the distinctness and obviousness of the points that he is called on to observe, it seems probable that this study calls forth (especially in young boys) a more concentrated exercise of the faculties it does develop than any other could easily do. If both the classical languages were to cease to be taught in early education, valuable machinery would, I think, be lost, for which it would be somewhat difficult to provide a perfect substitute.' (*Essays on a Liberal Education*, p. 133.)

The materials here spoken of must mean the subject matter of the ancient authors, and not simply the languages; this, however, does not help the case, as the matter can be far better given in translations. The second reason—the exclusion of the senses, and the simplicity and definiteness of the classification to be applied—must refer to the language part; but it contains nothing special to the classical languages. Moreover, as regards putting before the mind of a student distinct issues, and still more in adapting these to the state of his faculties and advancement, the learning of a language seems to me far inferior to most other exercises.

IV. *A Knowledge of the Classics is the best preparation for the Mother Tongue.*

This must have reference either (1) to the Vocables of the Language, or (2) to the Grammar and Structure of our composition.

(1) As regards the vocables, we have to deal with the presence of Latin and Greek words in English. There being several thousands of our words obtained directly or indirectly from the Latin, it may be supposed that we should go direct to the fountain head, and learn the meanings in the parent language. But why may not we learn them exactly as they occur in the mother tongue? What economy is there in learning them in another place? The answer must be, with a qualification to be given presently, that the economy is all in favour of the first course. The reasons are plain. For one thing, if we learn the Latin words as they occur in English, we confine ourselves to those that have been actually transferred to English; whereas in learning Latin as a whole, we learn a great many words that have never been imported into our own language. The other reason is probably still stronger, namely, that the meanings of a great number of the words **have** greatly changed since their introduction into English; hence, if we go back to the sources, we have a double task; we first learn the meaning in the original, and next the change of meaning that followed the appropriation of the word by ourselves. The meaning of 'servant' is easiest arrived at, by observing the use of the word among ourselves, and by neglecting its Latin origin; if we are to be informed

what 'servus' meant in Latin, we must learn further
that such is *not* the present meaning; so that the directing of our attention to the original, although a legitimate
and interesting effort, does not pertain to the right use
of our own language.

Besides the vast body of Latin words entering into
our language, as a co-equal factor with the Teutonic element, there is a sprinkling of special terms both Latin
and Greek, adopted for technical and scientific uses.
The appropriation of many of these is recent, and the
process is still going on.   Even with these, however, it
is unsafe to refer to the original tongues for the meaning;
we must still see what they mean as at present applied.
A knowledge of Greek would be a fair clue to the
meaning of 'thermometer,' and 'photometer,' and a few
others; but for the vast mass of these appropriations, it
gives no clue whatever, or else it puts us on the wrong
scent.   'Barometer,' as 'weight-measure,' would be most
suitably applied to the common beam and scales; the
real meaning would never be guessed. So, 'eudiometer'
cannot. suggest its meaning to a Greek scholar; 'hippo-
potamus' is equally enigmatic.   Of the 'ologies' very
few correspond to their derivation.   We have such conflicting names as 'astrology, astronomy;' 'phrenology,
psychology'; 'geology, geography,' 'logic, logographer,
logomachy'; 'theology, theogony'; 'aerostatics, pneumatics.' 'Theology' being the science of 'God,' 'philology' should be the science of 'friendship' or the
affections.   It was remarked by Mr. Lowe that the
word 'aneurism,' to a Greek scholar, would be misleading; he would not at once suppose that it is a
derivative of the Greek verb ἀνευρύνω, 'to widen.' So
26

with the word methodist,' the knowledge of Greek is not a help but a snare.

It is well understood to be a reason for borrowing foreign words, that they do not suggest any meaning but the one intended to be coupled with them. In obtaining words for new general ideas, our native terms contain misleading associations; the great virtue of the names— 'Chemistry,' 'Algebra,' 'rheumatism,' 'hydrated,' 'artery,' 'colloid'—is that we do not know what they originally meant; any designation that we could invent in our own language for such vast sciences as Chemistry and Algebra would contain some narrow and inadequate conception which would be a perpetual stumbling-block to the learner.

The only qualification to the principle of learning the meanings of words from present use solely, is, that the classical words in our language are mostly derivatives from a small number of roots; so that a knowledge of the meanings of say a hundred roots assists in discovering the meanings of thousands of derivatives. Not but that we must still check every derivative by present use; yet the memory is considerably assisted by a knowledge of the primitive meaning as partly retained in the numerous compounds. We must observe the present employment of the words—'agent,' 'actor,' 'enact,' 'action,' 'transaction;' nevertheless, when we are informed of the original sense of the root 'ago,' we are enabled thereby to obtain a speedier hold of the meanings of the derivations. So with the Greek roots,—'logos,' 'nomos,' 'metron,' 'zoon,' 'theos,' &c. This advantage, however, is attainable without entering upon a course of classical study. The roots actually employed in the language are separated and presented apart, and their derivatives set forth; and we are thus taught exactly that portion of the Latin and Greek vocabulary that serves the end in view.

(2) The argument as applied to the Grammar or Syntax of our own language is equally at fault. The natural course in learning the grammatical order of English sentences is to study and practise English composition. To be habituated to different sentence arrangements must be rather obstructive than otherwise. The reference to any other language can only be a matter of curiosity. If it ever happened that our language could borrow an effective arrangement of syntax from any other language, the borrowing should have taken place once for all, so that all succeeding ages might adopt it as a naturalized usage.

In connection with this argument may be taken the frequent allegation that the classics are an introduction to general Literature, as affording the best models of taste and style; in studying which we improve our compositions in our own language. There is here a host of loose assumptions. The excellence of the ancient writers is not uniform, and some assistance must be given to the pupil in discriminating the merits from the defects, a lesson that would be best begun in our own language. Moreover, the remark just made applies again. Whatever effects can be transferred by us to our own compositions cannot remain to be transferred now. The vast series of classical scholars that have written in the modern languages ought long before this time to have embodied whatever beauties can be passed on from the ancient literatures. In modern European literature, there is a large school of imitators of the ancient authors, through whom we can derive at second hand all the characteristic effects possible to be reproduced in modern compositions.

V. *The Classical Languages are an introduction to Philology.*

This argument is one of the recently discovered makeweights on the side of classical teaching. The science of Philology is a new science; and before launching it into the present controversy, its claims as a branch of school or college education should be established on independent grounds. Having its ultimate roots in the human mind, like a great many other sciences, it is a recondite branch of the vast subject of Sociology, or Society, viewed both as structure and as history. Its immediate sources are the existing languages of mankind, which are made the subject of comparative study, with a view to trace community as well as diversity of structure (whence springs *Universal Grammar*), and also historical connection and derivation. Such a subject may enter into the curriculum of the higher education, but not at a very early stage; it must allow priority to the more fundamental sciences.

Assuming that the subject is to be received among school and college subjects, the bearing of the classical languages is somewhat insignificant. Latin and Greek, as usually taught, are both defective and redundant in their bearing on General Philology. They are only two languages out of a multitude that have to be more or less minutely compared. The examples taken from other languages, Sanscrit for example, are of as great importance as those from Greek and Latin, and we cannot be expected to make an equal study of all these languages. In point of fact, we must be taught Philology by examples cited from many languages, which we do not pay any further attention to; and the Greek and Latin examples may be obtained in the same partial way. The full knowledge of the Greek and Latin authors does not avail us for this subject.[1]

These are the leading arguments in favour of the present system of classical study. The supposition is that by their cumulative effect they justify the continuance of the system after the original occasion of its introduction has ceased. On reviewing the tenor of these arguments, however, we find that, after all, they do not support the real contention; which is, that Latin and Greek, and they alone, as an undivided couple, shall continue to form the staple of our higher education. Several of the arguments apply equally to modern languages, and others would be met by the retention of Latin, by itself.

The case is not complete until we view the arguments on the other side.

---

[1] Mr. Sidgwick has some admirable remarks on this point in his Essay already referred to (p. 94). Mr. A. H. Sayce expresses himself strongly as to the small linguistic value of the two classical tongues. 'For purely philological purposes they are of less interest than many a savage jargon, the name of which is almost unknown, and certainly than those spoken languages of modern Europe whose life and growth can be watched like that of the living organism, and whose phrenology can be studied at first hand.' 'The greater the literary perfection of a language, the less is its importance to the mere glottologist' (*Nature*, November 23, 1876).

I. *The Cost.*

The amount of time consumed in classical teaching during the best years of youth is well known to be very great, although not everywhere the same. In most classical schools in this country more than half the time of the pupils is occupied with Latin and Greek for a number of years; and not long ago, nearly the whole time was taken up in many of our seminaries. In Germany, at the Gymnasia, six hours a week are given to Latin, for four years, and seven hours a week for other two years (age from twelve to eighteen): seven hours a week are given to Greek, for two years, and six hours a week for other two years (age from fourteen to eighteen). At the University, it is optional to pursue Classics.

The question, therefore, arises—Are the benefits commensurate with this enormous expenditure of time and strength? We might grant that a small portion of time —two or three hours a week, for one or two years— might possibly be repaid by the advantages; but we are utterly unable to concede the equivalence of the results to the actual outlay.

In the more recent system of teaching, under which some attention is given to the history and the institutions of Greece and Rome, a certain amount of valuable knowledge is intermixed with the useless parts of the teaching; and for this a small figure must be entered on the credit side. But all such knowledge could be imparted in a mere fraction of the time given to the languages.

The classical system has been the practical exclusion of all other studies from the secondary or grammar schools. For a long time, the only subject tolerated in addition was a very elementary portion of Mathematics —Euclid and a little Algebra. The pressure of opinion has compelled the introduction of new branches—·as English, Modern Languages, and Physical Sciences; but either these are little more than a formality, or the pupils are subjected to a crushing burden of distracting studies. To be in school five hours a day, with two or three hours for home tasks, is too great a strain on youths between ten and sixteen. Moreover, in the evening preparations, it is found that the classical lessons absorb the greater part of the attention.[1]

The argument from disproportionate cost is sometimes met by alleging the defectiveness of the usual methods of teaching the languages; and many short and easy methods have been propounded. Experience has not yet shown any means of seriously reducing labour; and the thing is not likely. A vast acquisition is unavoidably involved in any cultivated language. The Grammar and the Vocabulary cannot be committed to memory without a large expenditure of strength; and the authors to be read have each their special peculiarities to be mastered. The observance of the methods of good teaching will make a considerable and important dif-

---

[1] We are rapidly approaching a compromise between the new and the old systems, on the basis of omitting one of the two classical tongues, that is, Greek; the Latin alone to continue as an imperative branch of the curriculum of higher education. A considerable relief will no doubt be experienced by throwing Greek into option; but the radical evil of our Grammar School system will remain. The two best hours of the day for several years will still be given to a barren occupation; and the thorough reconstruction of the scheme of liberal studies will be indefinitely postponed.

ference, but will not dispense with the demand of two or
three hours a day for several years to attain a moderate
proficiency in Latin and Greek. Moreover, the system
as practised, throws away the best known device for
accelerating lingual study; namely, allowing a familiarity
with the subject matter of the several authors to be
attained in advance. The pupils in the Latin and Greek
classes have not as yet been initiated into any important
subject; and what renders the study tolerable is the
large devotion of time to the one theme of universal
interest—personal narrative.

II. *The mixture of conflicting studies impedes the course
of the learner.*

On the supposition that the classical languages are
taught, not in their simple character as languages, but
with a view to logical training, training in English,
literary culture, general philology,—the carrying out of
so many applications at one time, and in one connection,
is fatal to progress in any. Although the languages
may never actually be used, the linguistic difficulties of
the acquisition must be encountered all the same; and
the attention of the pupil must be engrossed in the first
instance with overcoming these difficulties. It is, therefore, an obvious mistake in teaching method to awaken
the mind to other topics and considerations, while the
first point has not been reached. I have everywhere
maintained as a first principle of the economy or conduct
of the Understanding, that separate subjects should be
made separate lessons. This is not easy when two
studies are embodied in the same composition, as lan

guage and meaning; in that case the separation can be effected only by keeping one of the two in the background throughout each lesson.

The least questionable effect of classical study (although one equally arising from modern languages) is the exercise of composing in our own language through translation. Still, it is but a divided attention that we can give to the exercise. We are under the strain of divining the meaning of the original, and cannot give much thought to the best mode of rendering it in our own language. This is necessarily a varying position. There may be occasions when the sense of the original is got without trouble, and when we are free to apply ourselves to the expression—in English, or whatever language we are using. But this is all a matter of chance; and such desultory fits of consideration are not the way to make progress in a vast study. Moreover, the master is a man chosen because he is a proficient in classics, not because he has any special or distinguishing acquaintance with the modern tongue. Now it must seem incontestable that the only way to overtake an extensive and difficult department of information and training, is to proceed methodically, and with exclusive devotion of mind at stated times, under the guidance of an expert in the department. All experience shows that only very inferior English composition is the result of translating from Latin or Greek into English. There is necessarily a good deal of straining to make the English fit the original; while the greater number of the most useful forms of the language are never brought into requisition at all.

There is something plausible in the supposition of cultivating all the faculties at one stroke, as if an exercise could be invented that could teach spelling, cooking, and dancing, simultaneously. Because the same piece of composition involves grammar, rhetoric, scientific information and logical method, we are not to infer that it should be the text for all these lessons at one time. It is not merely that the way to carry the mind forward in the several departments is, to keep it continuously fixed on each for a certain duration; equally pertinent is the fact that, although every passage occurring in a lesson must needs embody language, rhetoric, and information, the same passage does not equally suit for all the applications.

It may be true that classical education is many-sided ; but what if it is defective on each side ? 'The very fact that the same instrument is made to serve various educational purposes, which seems at first sight a very plausible argument in its favour, is really, for the majority of boys, a serious disadvantage.' (Sidgwick, *ut supra*, p. 127.)

The study of fine Literary effects cannot be carried on in connection with Latin and Greek, not only because of the distraction of the mind with other things, but because of the random, uncertain, unconsecutive way that the examples are brought forward. Even if there were no order whatever in the parts of a subject, still the irregular presentation of these would be adverse to a cumulative impression. The same would apply to General Philology, if that were regarded as one of the uses of classical study.

The conclusion on the whole is, that the teaching of language is most rationally conducted when it stands on the original footing of the classical languages in the fifteenth and sixteenth centuries, *i.e.* when the language itself, as a means of interpretation and communication, is the fact, and the whole fact. The attention of the pupils could then be kept to the one point of mastering grammar and vocables: the authors studied would be studied with this sole aim. The language teacher is not an interpreter and expounder of history, poetry, oratory and philosophy, but an instrument for enabling the pupils to extract these from their original sources in some foreign tongue.

III. *The Study is devoid of interest.*

This may not be universally admitted, but it is sufficiently attested for the purposes of the present argument. There is, first, the dryness inseparable from the learning of a language, especially at the commencement. There is, next, the circumstance that the literary interest in the authors is not felt, for want of due preparation. It is a fact that, but for the never-failing resource of sensation narrative, by which we arouse the dormant intellect of the child in the second standard, the reading of classical authors would be intolerable at the early age when they are entered upon.

It is the nature of science to be more or less dry; until its commanding power is felt the path of the learner is thorny. But literature is nothing, if not interesting. There should be even in a course of Belles-Lettres, a certain amount of science, in the shape of generalities and technicalities; but these are soon passed, and the mind is free to expatiate in the rich pastures of the literary domain. Literature, instead of being the dismal part of the school exercises, should be the alternative and relief from Mathematics and the elements of Science generally. This cannot be, if the pupils are thrust prematurely upon a foreign literature while mastering several new vocabularies. It is now plain to the best educationists, that our own literature must be the first to awaken literary interest, and prepare the way for universal literature.

IV. *The study panders too much to authority in matters of opinion.*

The classical student is unduly impressed with the views promulgated by the Greek and Roman authors, from the very length of time that he is occupied with them. The authority of Aristotle, once paramount in the world of thought, has long ceased to be infallible, but the reference to his supposed opinions is still out of proportion to any value that can now belong to them. Any views of his as to the best form of government, as to happiness and duty, are interesting as information, but useless as practice.

A curious and expressive incident occurred at a recent meeting of the British Association. Sir William Thomson, in the course of a paper read before his section, desired his hearers, when they went to their homes, to draw their pens through a certain paper of his in their copies of the 'Proceedings of the Royal Society.' It would be well if the example were imitated by every philosopher that has happened to change any of his opinions. Even if we accorded to Aristotle a command-ing sagacity in Ethics and in Politics, we should like to have his latest decisions as to the value of what we now possess as his writings.

Note on recent views of the Classical question.

Mr. Henry Sidgwick.—The article that has been repeatedly referred to concludes with the following recommendations as to the subjects suitable for Higher School education.

'1 think that a course of instruction in our own language and literature, and a course of instruction in natural science, ought to form recognised and substantive parts of our school system.' 'I think also that more stress ought to be laid on the study of French.' To make room for these additions, the obvious remedy is 'to exclude Greek from the curriculum, at least in its earlier stage.' 'It is supposed that there is a saving of time in beginning the study of Greek early. I am inclined to think that very much the reverse is the case, and that, if several languages have to be learnt, much time is gained by untying the faggot and breaking them separately. There are two classes for whom the present system of education is more or less natural, —the clergy, and persons with a literary bias and the prospect of sufficient leisure to indulge it amply. Boys with such prospects, and a previous training of the kind I advocate, would in the average feel, as they approached the last stage of their school life, an interest in Greek strong enough to make them take to it very rapidly.' 'The advantage that young children have over young men in catching a spoken language, has led some to infer that they have an equal superiority in learning to read a language that they do not hear spoken; an inference which, I think, is contrary to experience.'

Mr. Alexander J. Ellis.—In a Lecture, on the Acquisition of Languages, delivered before the College of Preceptors, Mr. Ellis criticizes severely the English School system. He remarks on the absurdity of talking of the humanizing effect of the Latin and Greek languages, of the grand literatures they contain, and so on—when the one condition is wanting, namely, 'that those who acquire them should be able to use them.' 'The tree of language is indeed vast in our schools; but it is, after all, but an overgrown weed. Good masters learn to hang many a garland on its unsightly knots by the way, and to bend many of its branches into unnatural but more or less useful directions.' 'These results are not legitimate deductions from teaching languages.'

'Every speaker is bound to know his own language first, without relation to other languages.' 'Lessons in language should be supplemented with lessons on things. We must have something to speak and write about besides language itself.' After English, should come German and French. 'Hitherto, German and French have been regarded as the accomplishments, and Latin and Greek as the staple, of literary education. It is time to reverse the terms. Latin and Greek have drifted into being accomplishments.' 'If a boy is "up" in English at ten; knows his German, to the extent indicated at twelve, and his French at fourteen; he will be a better Latin scholar at sixteen, and Greek scholar at eighteen, than the majority of those who leave our public schools.' 'Literature is one of the very last things to be attacked. To appreciate it, requires much education, often much experience of life, and great familiarity with the language, and often with social habits and customs.'

Mr. Matthew Arnold.—At the close of his Report to 'The Schools' Inquiry Commission' on the Middle Schools of Germany, Mr. Arnold adverts to the conflict of the modern spirit with the old exclusive classical system, and indicates what he considers the true solution. 'The ideal of a general, liberal training is, to carry us to a knowledge of ourselves and the world. We are called to this knowledge by special aptitudes which are born with us: the grand thing in teaching is to have faith that some aptitudes for this everyone has. This one's special aptitudes are for knowing men—the study of the humanities; that one's special aptitudes are for knowing the world—the study of nature. The circle of knowledge comprehends both, and we should all have some notion, at any rate, of the whole circle of knowledge.

The rejection of the humanities by the realists, the rejection of the study of nature by the humanists, are alike ignorant. He whose aptitudes carry him to the study of nature should have some notion of the humanities; he whose aptitudes carry him to the humanities should have some notion of the phenomena and laws of nature. Evidently, therefore, the beginnings of a liberal culture should be the same for both. The mother tongue, the elements of Latin and of the chief modern languages, the elements of history, of arithmetic and geometry, of geography, and of the knowledge of nature, should be the studies of the lower classes in all secondary schools, and should be the same for all boys at this stage. So far, therefore, there is no reason for a division of schools. But then comes a bifurcation, according to the boy's aptitudes and aims. Either the study of the humanities or the study of nature is henceforth to be the predominating part of his instruction.'

CHAPTER XI

THE RENOVATED CURRICULUM.

ON the supposition that Languages are in no sense the main part of Education, but only helps or adjuncts under definite circumstances, the inference seems to be, that they should not, as at present, occupy a central or lead ing position, but stand apart as side subjects available to those that require them.

I conceive that the curriculum of Secondary or Higher Education should, from first to last, have for its staple the various branches of knowledge culture, including our own language. The principal part of each day should be devoted to these subjects; while there should be a certain amount of spare time to devote to languages and other branches that are not required of all, but may be suitable to the circumstances of individuals.

The essentials of a curriculum of the Higher Education may be summed up under three heads:—

I. SCIENCE, including the Primary Sciences, as already set forth; some one or more of the Natural History Sciences—Mineralogy, Botany, Zoology, Geology; to which may be added Geography. To what extent this vast course should enter into general education has already been sufficiently discussed. Our present purpose does not require the nice adjustment of details.

II. A course of the HUMANITIES, under which I include (1) History, and the various branches of Social Science that can be conveniently embraced in a methodical course. Mere narrative History would be merged in the Science of Government, and of Social Institutions, to which could be added Political Economy, and, if thought fit, an outline of Jurisprudence or Law. This would put in the proper place, and in the most advantageous order of study, one large department recently incorporated with the teaching of the classical languages. by way of redeeming their infertility.

(2) Under the Humanities might next be included a view, more or less full, of Universal Literature. Presupposing those explanations of the Literary Qualities and Arts of Style that should be associated, in the first instance, with our own language, and also some familiarity with our own Literature, we could proceed to survey the course and development of the Literature of the World through its principal streams, including of necessity the Greek and Roman Classics. It is needless to add that this should be done without demanding a study of the original languages. How far a Philosophy of Literature should penetrate the survey I do not at present enquire. Materials already exist in abundance for such a course. It is the beau-ideal of Rhetoric and Belles-Lettres as conceived by the chief modern authorities in the department, as for example, Campbell and Blair in last century. Only, I should propose that the elements of Rhetoric, in connection with our own Literature, should lead the way.

Such a course would carry out, with effect and

27

thoroughness, what is very imperfectly attempted in conjunction with the present classical teaching. A tolerably complete survey of the chief authors of Greece and Rome, with studies upon select portions of the most important, could be achieved in the first instance; and it might be possible to include also a profitable acquaintance with the great modern literatures.

III. ENGLISH COMPOSITION AND LITERATURE. — This might either pervade the entire curriculum, or be concentrated in the earlier portions, the General Literature being deferred. What it comprises, according to my view, has been sufficiently stated. The survey of Universal Literature, would operate beneficially upon the comprehension of our own.

These three departments appear to me to have the best claims to be called a Liberal Education. The deviation from the received views is more in form than in substance. I would not call Science alone a Liberal Education, although a course that implied a fair knowledge of the Primary Sciences, a certain amount of Natural Science, and a wide grasp of Sociology, would be no mean equipment for the battle of life. I think, however, that the materials of Sociology might be accumulating all through the curriculum, and might serve to alleviate the severity of the strictly scientific course.

I think, moreover, that a Liberal Education would not be generally considered complete without Literature, although people must needs differ as to the amount. I hold that the three departments stated are sufficiently comprehensive for all the purposes of a general educa-

tion, and that no other should be exacted as a condition of the University Degree—the received mode of
stamping an educated man.

Such a course should be so conducted as to leave a
portion of time and strength for additional subjects. An
average of two to three hours a day might be occupied
with the continuous teaching in the three departments.
Assuming a six years' curriculum—covering the Secondary School and the University courses—it is easy to see
that a large amount of thorough instruction might be imparted in those limits; leaving perhaps one third of the
pupils' available time, for other things.

Of the extra, or additional subjects, Languages would
have the first claim. These, however, should not be
under any authoritative prescription; they should never
enter into any examinations for testing general acquirements. Every person going through such a course as
we have supposed, would be urged and advised to take
up at least one foreign language, giving the preference
to a modern language: the intention being to learn
it up to the point of use as a language. How many
languages any given person should study must depend
upon circumstances. The labour of a new language
is not to be encountered without a distinct reason. It
is never too late to learn any language that we discover
ourselves to be in want of. If we need it for information
on a particular subject, we can learn it up to that point
and no farther.

An hour every day may be available at any part of
the course for a new language, whether modern or
ancient. If either Latin or Greek is taken up, it would
be learnt strictly by the grammar and the dictionary;

just as Dutch or Gaelic would be learnt: we should not diverge into literary matters, or the criticism of beauties; all which would be reduced to a small compass, after a survey of the literature, and a familiarity with good translations.

There would be no need to begin the study of language early, and little advantage: and it would be undesirable to take two languages together. There are other matters to divide the extra hours with languages. I need only mention Elocution as appertaining to every one. For more special tastes would be provided Music and Drawing. There would also be a variety of special courses on branches of knowledge not embraced in the regular curriculum. In a well-provided institution, there might be classes devoted to Anglo-Saxon, General Philology, select portions of History, and so on. I am not specially adverting to the topics preparatory to the several professions.

The reasons for the change now proposed have been given in substance already. They are contained in the general argument as to the position of languages in general, and of classics in particular. Besides the consideration that languages should be learnt only when meant to be used as languages, I have all along put great stress on the wastefulness of carrying on several incongruous lessons at one time. From the first statement of the Laws of Agreement onwards, I have contended for the necessity of like going with like in the same exercise.

I have also urged the economy of learning language after laying up a good stock of ideas. Setting aside the pronunciation of a foreign language, the acquisition of the grammar and the vocabulary is easier late than early; any decay in the plastic force of memory is more than made up by the other advantages.

The scheme thus set forth appears the only means of arresting the tendency inevitable at the present day to excessive specializing of the studies constituting a liberal education. It is the supposed necessity of retaining dead languages and of adopting foreign living languages as an integral part of education, that leads to options so very wide as to leave out science almost entirely from one course, and literature almost entirely from another. A mere language course, containing as it does irregular smatterings of history and of literature, is not an adequate cultivation of the human faculties; it is defective both on the side of training and on the side of knowledge imparted. On the other hand, I regard it as equally undesirable to limit the course of study to science, still less to physical science (excluding Logic and Psychology), least of all to Mathematics and Physics.

The more obvious objections to the proposed curriculum may be glanced at.

First. It will be called by the dreaded name—Revolution. Yet the revolutionary element is not very great after all. It consists only in putting languages in the second place, reserving the first to the subject-matter. The scheme pays great regard to the element of the antique, as represented by Greece and Rome, and would render the acquaintance with the history and literatures of both countries, more general and more thorough than at present. A day may come when this amount of attention will be thought too much.

Second. Classics will be 'ruined. To this there are several answers. According as people believe the classical languages to be useful, they will keep them up to that extent and no more. But classics will never cease, so long as the existing endowments continue. A small number of persons will always be encouraged to master those languages thoroughly, so as to maintain the study of the history and literature of the ancient world. The teachers of ancient literature would be expected to know the originals; and they alone would constitute a considerable body.

Third. Some minds are incapable of science, and more especially of Mathematics, the foundation of the whole. In answer to this we may freely concede, that many minds find abstract notions exceedingly distasteful and, as a consequence, difficult. Men of admitted ability have been found incapable of mastering Euclid, while at home in languages, and in literature. In this case, however, the disproportionate pursuit of the one department has been the real obstacle. The experience of existing Universities shows that four men out of five can pass for a degree, containing elementary Mathematics. Perhaps their comprehension of the subject is not great or exact; but if their minds were more disengaged, they could understand it sufficiently to go on with a course of the experimental and other sciences, in which the interest would be more universal.

Although there are men of good judgment or practical sense, who have never had any abstract teaching, and might seem incapable of it, yet the highest order of judgment combines both abstract notions with concrete experience; and in a thoroughly liberal education, abstract science ought not to be dispensed with.

It may be remarked finally that any man possessing a thoroughly grammatical knowledge of several languages is not wanting in aptitude for abstract science; grammar does not amount to a scientific discipline, but it attests the capability of undergoing such a discipline.[1]

[1] The curriculum now roughly sketched would harmonize the course of primary and secondary education, and do away with the troublesome bifurcation of the Ancient and the Modern sides, which at present complicates and embarrasses our higher schools and colleges. The work of the primary school is necessarily on the lines here laid down, and could be made still more profitable by a closer adherence to the same plan. There would be a common ground for all the professions to meet.

CHAPTER XII

THE difficulties of moral teaching exceed in every way the difficulties of intellectual teaching. The method of proceeding is hampered by so many conditions, that it barely admits of precise demonstration or statement.

Morality is in the situation of the mother tongue: it does not depend solely on the school teacher, or on any one source ; it is imbibed from innumerable sources ; and the school does not even rank as one of the chief. There are unquestionably inborn tendencies, more or less powerful, to make men prudent, and just, and generous, when once they are placed in the suitable circumstances. But experience shows that these native forces are not fully adequate to the desired end ; and society super-adds a special discipline to make up for the defects. The greater part of this discipline, however, is not teaching, in the common meaning of the word, but the public dispensation of punishment and reward.

As man is not a solitary animal, but spends his whole life in the society of fellow-beings, there grows up in the individual bosom a set of social sentiments or feel-ings of a very mixed character ; the dependence of each one on the rest involves both the lowest and the highest of our gratifications. We have to think of others in everything that we do : our personal wishes are biassed by what those about us wish ; our conduct is shaped by our various social relationships.

What we have to do to others, and what we have to expect from others, are at first learnt by personal experience. We are introduced to society in a state of total dependence, we follow our own wills only in so far as we are allowed ; and we have to accommodate ourselves to our circumstances, to do and to refrain from doing, at the dictation of superior power. This habituation to obedience, in prescribed lines, is our first moral education, and represents by far the greatest part of that education in its whole compass. By acting and re-acting on the numerous individuals that we encounter in various social relationships, we obtain both the knowledge of duty, and the motive to do it.

Besides our own personal contact with parents, masters, superiors, friends, and the influence exercised by those on our own conduct, we are witnesses to the demeanour of our fellows in their dependence upon the surrounding society. We see the obstruction offered to their free-will, the pain inflicted on transgression, the approval of obedience and compliance. In short, we are taught by innumerable *examples* what society wills that each person should do, and what are the consequences of doing and of not doing. All these examples we take to heart, and they augment the influence of society in our moral education.

This primary and personal source of moral education is analogous to the education in physical laws by personal experience of their working for good and for evil. We learn to subject our conduct to the influences of the natural world—to avoid stumbling, running against posts, scalding ourselves, and drowning ; to court all pleasant things—sunshine, warmth, sweet tastes, nourishment. We are soon habituated to a set conduct in accordance with these physical laws, and that without anybody's teaching ; although we ultimately derive great benefit from the fruits of other people's experience.

It is quite conceivable that our personal intercourse with human beings in varied relationships might be of itself sufficient to give us all the moral habits necessary to a good citizen ; just as the children in a cultivated family acquire language and breeding, of the highest degree of polish, by mere unavoidable imitation. Indeed, if we survey the history of the human race, in the vast majority of situations no other education is given. The child learns to avoid blows, to conciliate favour, by its own encounters with parents, companions, superiors, and equals ; extended by observation of others doing the same. No other moral lessons are read in the usages of the uncivilised tribes. The situation is even reproduced among ourselves. The virtues of the soldier are formed almost exclusively by the operation of the army penal code. He knows by his own experience and by the observations of others, the penalty of disobedience ; he avoids that penalty first by a special volition, and subsequently by acquired bent or habit.

After clearly understanding that the mutual encounter of human beings socially related is the one never-failing source of social good conduct, in other words, morality (in its beginning and its type), we can consider what are the defects or shortcomings of this method, and by what other methods these are found to be best overcome. In such subsidiary devices consists what is usually styled moral teaching, which is the corrective of the hard matter-of-fact teaching of good and evil consequences, just as the science and traditions of the race improve upon the individual experience of the physical laws.

Whatever may be the supplementary modes of enforcing morality, we may assume that they are in keeping with the primitive, the sure and perennial mode, of trial and error, or actual experience of the human relations. The rude and painful shock of collision with the wills of others may be anticipated and thrust aside in such a way that the moral may be still impressed ; or when actually occurring, it may be so improved upon by a well-managed commentary, as to be avoided in the future. But in either case, the motive power is what happens in real life ; the evil and the good that we experience at the hands of others are the forces for keeping us in the orbit of duty.

The schoolmaster, in common with all persons exercising control for a particular purpose, is a moral teacher or disciplinarian ; contributing his part to impress good and evil consequences in connection with conduct. For his own ends, he has to regulate the actions of his pupils, to approve and disapprove of what they do as social beings related to one another and to himself. He enforces and cultivates obedience, punctuality, truthfulness, fair dealing, courteous and considerate behaviour, and whatever else belongs to the working of the school. Whoever is able to maintain the order and discipline necessary to merely intellectual or knowledge teaching, will leave upon the minds of his pupils genuine moral impressions, without even proposing that as an end. If the teacher has the consummation of tact that makes the pupils to any degree in love with the work, so as to make them submit with cheerful and willing minds to all the needful restraints, and to render them on the whole well-disposed to himself and to each other, he is a moral instructor of a high order, whether he means it or not.

This, however, is not all that is expected of the ordinary teacher—at least in the primary schools. He (or she) is expected to give express lessons of a moral kind, whether with, or apart from, religious lessons ; and these lessons are something over and above what grows out of the work of teaching knowledge elements. The teacher is assumed to be something more than one of our fellow-beings echoing approbation and disapprobation, and swelling the chorus of voices that engrain right dispositions on the youthful mind. As an intellectual or scientific expositor, probably also as a persuasive monitor, he concentrates and methodizes the scattered and random moral impressions of every-day life, so that ' a day in his courts is better than a thousand' in the general world.

The additional moral teaching by separate lessons, having no reference to the actual incidents of the school, must operate by referring to ideal incidents and situations, chosen for their illustrative character. The recollection of actual facts, and the imagination of depicted facts, are appealed to, and their moral lesson duly expressed. This exercise has its advantages and its disadvantages.

The advantages resemble the superiority of experi-

· ment to observation in science.   Cases in point are contrived to show the evils of the various vices, and the good consequences of the virtues.   A cumulative impression is made in favour of the line of conduct that ought to be followed in given situations.   The exposition of the mischiefs and the dangers of falsehood, instead of being left to chance occurrences and scattered effects, here a little and there a little, is made more emphatic by gathering together a host of instances, real and feigned, working to one total effect.

For such lessons, a good classification of virtues and vices is a prime essential.   The teacher needs to have a clear scheme before him, in order to concentrate his teaching.   If the same thing is repeated under various names, the result is mere distraction of mind.   The fundamental virtues need to be grasped in the first instance, and to be indicated under their best-known designations : also they should be exemplified in pure and typical instances.   The mixed and modified virtues are then rendered intelligible.

The main disadvantage of the scheme arises from the weak conceptive power of the pupils.   Imagined cases of virtue and vice do not always have their full effect, with minds that are little experienced in the ways of the world.   It becomes necessary to put the examples in forms that are too unqualified, and that leave defective and one-sided impressions, not easy to be got rid of.

In Ethics, as in other subjects, there may be a desul·tory treatment, preparatory to the regular and methodical treatment.   Instances may occur conveniently by chance, and may be used to make an impression ; but then, like cases in law, they must carry their principle on their face, which requires them to be properly generalized, and this involves the same subtle perception on the part of the pupil as is implied in the understanding of the classified virtues.

A few words on the Classification of the Virtues. The cardinal virtues, in the modern treatment, are Prudence, Probity or Justice, and Benevolence. PRUDENCE is sometimes described as the Duties that we owe to ourselves, but this is not the most suitable expression. Prudence, or self-regarding conduct, stands in a very different position from the two other cardinal virtues: it has the support of our own natural self-seeking impulses. The obstacles to be overcome are—want of knowledge, and present impulse. Knowledge is gained in time, and may be aided by teaching; impulse can be to a small extent checked or controlled by guidance, admonition, and representation of future consequences. The important point is, that this is not the region of authority, except in the parental sphere; but the region of friendly advice, information, and assistance. It is the more necessary to attend with rigour to the speciality of the prudential virtue, that we are always prone to assume the air of authority in dealing with those that are in our power: and moreover, there is an easy pretext for making prudence a matter of obligation, inasmuch as, if any man is imprudent in his own affairs, he is likely to fail in some of his duties to others. If a parent is idle, spendthrift, or drunken, his family suffer. Nevertheless, it will be found that there is a gain in persuasiveness, by taking each virtue in its own proper character, in the first instance; and the proper character of prudence is the enlightened regard to our own interest. This is the first and easiest conquest over our inherent moral weaknesses. The line to be pursued is special and distinct.

The aspects and departments of Prudence—as Industry, Thrift, Temperance—are all intelligible, and should be kept in view by the teacher, in his scheme of the virtues.

A very great part of Prudence unavoidably concerns our relations with others ; for to get the most we can from life, we must behave well to everyone that has the power to help or to thwart us. This social situation also brings into view our duties properly so-called—Justice and Benevolence ; still we must rigidly abstain from entering on these, while our aim is to impress the self-regarding proprieties. The reasons will presently be seen.

The virtue of Probity, or JUSTICE, ranks first among our social duties or obligations. Justice is the protection of one man against every other man ; it is what is embodied in the laws, and enforced by penalties. The promulgation of these penalties, as already remarked, is the primary teaching of Justice. The teaching of the schoolmaster co-operates, by endeavouring to correct in advance the evil dispositions that incur the penalties. The essential idea of Justice is reciprocal good, and reciprocal abstinence from harm. It is the conduct imposed alike upon all, for the advantage of each. Nobody is expected to do more or less than what is prescribed for every member of the society.

The virtue of BENEVOLENCE is something beyond justice. It is doing good irrespective of the social necessities that Justice proceeds upon. It is not enjoined by penalties, but recommended to the voluntary choice of individuals. Its chief occasion is distress or privation arising through inequality of fortune, and through the accidents that render individuals unable to support themselves.

Self-sacrifice, devotedness, kindness, pity, compassion, doing good, beneficence, philanthropy—are among the numerous designations for this portion of moral duty. Besides all which, there are certain qualities that seem either to fall under the two other heads or to stand altogether apart, but that really come under the present head. Fortitude, courage, constancy, contentment—are prudential virtues to appearance, but the high praise accorded to them shows that they are supports of Justice and Benevolence ; Honesty is a name for probity carried to the length of positive benevolence.

The virtue of Truth is sometimes regarded as an independent virtue ; but, in reality, it is an adjunct of the others. It is a remarkably precise virtue : it does not admit of gradations, in the same sense as the others; it is a matter of yea or nay.

These three fundamental virtues cross and re-cross at so many points, that it needs a steady grasp to hold each class firmly according to its essential nature ; yet this is what the moral teacher should be able to do, if he is to marshal his resources in the most effective way. A good course of Moral Science should impart this fundamental qualification.

Next to classifying the Virtues, is the correct apprehension of the Motives. There is an equal liability to confusion as regards these. The fundamental division is into Self-regarding and Extra-regarding or Social ; and as each of these classes readily simulates the other, there is the same necessity for viewing each in its pro. per character at the outset. Prudence is the area of self-regarding motives : Justice supposes a mixture of the self-regarding and the social : Benevolence is the region of the Social or altruistic regards, pure and proper, together with a certain high and refined class of self-regarding motives, growing out of our social dispositions.

The appeal to the SELF-REGARDING motives follows a line peculiar to itself, which is well understood in oratory. It consists in making apparent the bearings of conduct upon the individual's own welfare; and is to be kept distinct. The virtues of industry, thrift, temperance, devotion to study or to knowledge, have each their reward, which ought to be rendered as palpable and evident as possible. All this is to be clearly distinguished from the social bearings, in order that each one of the forces may attain its full momentum. Moreover, there are many reasons why it is much easier to work upon the selfish feelings of men, than upon the other class.

It is in addressing the SOCIAL MOTIVES that we are chiefly liable to commit mistakes. We are here working upon the exceptional part of the human constitution, the small corner of self-devotion ; and we are in constant danger of quitting the narrow road to it, for the broad way of prudential self-regard. We shall not succeed in evoking great virtues by teaching and persuasion, unless we can clearly keep before us the social motives, first in their purest type of absolute self-sacrifice, and

28

next in their mixed character as made up of the social cravings and pleasures.

There are many arts bearing on such an attempt, which are minutely detailed under the philosophy of Sympathy. The one point to be steadily kept in view is this. The social aptitudes, like everything else, must be exercised ; and the mode of exercising them, is by directing and securing the attention upon the wants and the feelings of others. The most palpable form is Pity for manifest distress ; next is Sympathy with the pleasures of our fellow-beings ; and, by plying these exercises, a habit of taking interest in those about us is likely to be fostered. So to conduct the operation as to keep out altogether the self-seeking motives, is the delicate part of the task.

The second-class motives of Sociability—the cravings for Love, Affection, Pity—are perhaps the most powerful instrument of moral suasion : for, while having a genuinely altruistic side, they contain a very large mass of purely self-regarding emotions. To urge them exclusively is to degenerate from the high standard of pure altruism: and the most successful result will not be anything very lofty. Yet there is a lower deep, and a greater danger, namely, to make the cultivation of mutual regards bring forth, not pure affection, but mere worldly advantage.

A third survey equally necessary for the moral teacher, is the Relationships of Society : beginning at the Family, and extending to the State and the World at large. A clear conception should be attained of the exact bearing of every one of the Social groupings —what it is intended for, and what it is not intended for—so as to define the conduct suited to each. This comes under the wide department of Sociology, or Social Science and Philosophy, which is irregularly provided for in the school curriculum. The study of the relationship of parent and child, master and servant, ruler and subject —has an understood moral bearing, and may be prosecuted in that view.

A further condition of successful moral teaching is a command of the stores of Language and diction that embrace the topics of moral suasion. This leads at once to the higher region of spoken address, as exemplified by our greatest orators. Without a certain compass of expression, and that well directed to the purpose, no one can hope to produce deep moral impressions by mere teaching ; and hence very little is to be expected of the common schoolmaster working in his own strength. It is only by being provided with good and suitable compositions, to be made use of in his teaching, that he can exert any influence raised above the effect of the common-place maxims floating in society—' Honesty is the best policy,' 'be just before you're generous,' 'in all things look to the end,' 'bear and forbear,' 'do all the good you can,' &c.

The briefest glance at moral teaching must not omit the topic of Moral Ideals. It is in Morality, more especially, that the teacher works by putting forward grand, lofty, and even unapproachable Ideals ; the supposition being that the charm and attractiveness of these will make a far more powerful impression than any unvarnished statement of consequences. From the earliest recorded ages, the moral education of mankind has proceeded upon a system of habitual exaggeration, as if the naked truth of things would not answer the end. Hardly any usage has a larger *consensus* than this. The miseries of vice and the glorious prospects of virtue are always depicted in terms far beyond the fact. Unless in some degree successful for its purpose the practice could hardly be so universal. The moral influence of an excited ideal of future good or evil must be looked upon as something very considerable. For we cannot be blind to the dangers or disadvantages of the system. In substituting the license of imagination for the restraints of truth, we incur serious liabilities. There must be some limits set to exaggeration, even for its own purpose: and these limits have not been duly observed in the stimulating compositions that are embodied in the lesson-books. But until this whole subject has been revised for wider application than the school, it is not to be expected that the teacher should quarrel with the materials put into his hand. All that he can do is to keep well before him the sober facts of life, as a counterpoise to the poetic flights of the lesson-book. While in the Ideal, self-devotion or self-sacrifice is depicted so as to kindle a momentary glow, the hard reality warns us that only a very small portion of this can be engrained in the average individual. Rivalry, competition, over-grasping and supplanting—are what we have to deal with on one side; and on the other, we have to set the tendencies to the social, the sympathetic, and the amiable; and close is the game we have to play in the encounter.

In the few remaining pages that can be devoted to this great subject, I will indicate what most requires caution in plying moral lessons.

1. A large part of the tactics of the teacher is determined by the natural repugnance of human nature to the whole subject. Pupils would much rather be instructed in knowledge than be lectured on virtue; while, as regards knowledge, want of liking is not so fatal to the end. The use of the fable, the parable, and the example, is evidently meant to avoid direct lecturing, and to reach the mind by insinuation and circumvention.

The lesson that arises unsought, that obtrudes itself on the attention when engaged in other matters, is the most effective of all. This is one of the incidents of historical reading ; but there should not be an obvious contrivance to bring it about. Moral reflections that are self-originating, that arise unavoidably in a given situation, are likely to yield an abiding impression. The spectacle of disaster from want of forethought, from quarrels and dissensions, from culpable ignorance, awakens the thoughtful mind to the value of the leading prudential virtues.

Tales, narratives, and biographies, that suggest the high moral qualities, produce their greatest effect when spontaneously perused ; their recurrence in the lesson-book, as tasks, introduces the drawback of dictation from without, which it needs much tact on the part of the master to suppress. The books of the children's own choice, read without any responsibility, are their most persuasive monitors.[1]

---

[1] 'The well-meant but futile '—' Hence we should learn,' and 'how im•

2. The moral teacher all through must work by con-ciliation, and not by fear. In order to produce a certain external appearance of good conduct, fear and punish-ment will succeed; but the inward sentiment cannot be gained in the same way. Wherever concealment is prac-ticable, we need to address the free-will of the subject. The attempt at compulsion only increases the natural per-versity. At an early age, when the child is pliable under influence, and not so much given to self-assertion, re-probation and punishment sink into the mind, and mould its inward sentiments of right and wrong ; but we must carefully watch the moment of transition from this humble and susceptible period to the development of egotistic preferences, under which the same system will no longer work. For boys and girls above twelve, we may as a rule pronounce that moral lecturing, except in actual discipline, is misplaced; and only a very round about approach to the subject can be borne. In the higher Schools and Universities direct moral teaching is by common consent disused as part of the ordinary class work.

3. We may appeal at any time to the prudential or self-regarding motives, provided we do not seem to be courting pretexts for repression and prohibition. If we

portant it is ever to remember,' answer no purpose whatever in Education, except that of giving the *congé* to the minds of children, whether as audi-tors or readers : it is a—' Now you may go, while I preach.' The efficacious mode of instilling moral principles, as suggested by the history of nations, is, at choice moments, and when all minds are seen to be in a state of gentle emotion, and in a plastic mood, to drop the word or two of practical in-ference, to enounce the single, pithy, well-digested sentiment, which, by its natural affinity with the excited feelings, at the moment, shall combine itself with the recollected facts. (ISAAC TAYLOR, *Home Education*, p. 258.)

show genuine anxiety for the interests of our pupils, we shall be listened to with attention; our chief danger is, that we look further ahead than their vision can follow us. To paint a picture of future consequences that shall be neither extravagant nor unintelligible is the standing difficulty. At an early age insurmountable, it diminishes with years, but always puts the utmost strain upon the tact of the teacher; nor are there many good models provided as aids.

The intermediate class of motives—neither purely self-regarding, nor purely heroic—are the social likings and affections, including compassion and pity. The cravings for affection undergo considerable changes in the critical periods of mental growth. At first, dependence inclines to the loving mood ; then comes the age of vigorous impulses and self-sufficiency, when the lust of power carried to domination and cruelty is rampant, —as in the flower of boyhood. Very little is now gained by appeals to love, affection, and pity ; the moment chosen must be very opportune to give a chance of success. Possibly this is the age when the higher altruistic or heroic motives may be used. Yet it is in the employment of these higher appeals that cautious reserve is most wanted. There is in every human breast a certain response to the trumpet call of heroic self-sacrifice ; but it should be kept for special occasions, and not wasted. The quantity of it that ever turns to action is very small in the mass of minds ; the earth is salted by the heroism of the few.

There is a mixed sentiment, containing a spice of the heroic, with a large element of the egotistic, that can be successfully appealed to in counterworking the baser forms of egotism.　This is the sentiment of Honour and Personal Dignity, which is cherished by social position, and may be found in all but the most worthless.　To stigmatize conduct as low, debasing, degrading, shame-ful, dishonourable, unworthy, is a very powerful weapon, at all times; and it is the appeal found most telling in the unruly years of youth.　The temperance orators have not discovered a stronger buttress to their cause than this.　It would be advantageous to every teacher to be able to wield this topic of address with skill and delicacy, care being taken to husband it for great emergencies.

4. Much as Plato has been criticized for his severe judgment on Poets, the fact remains that, taken as moral teachers, they are given to exaggeration.　They are artists first, and moralists next; and art, aiming at the agreeable, is adverse to imposing restraints or self-denial. When the sphere of Poetry is extended, as it ought to be, to include Romance, we feel at once the force of the observation.　The poet expresses, as no one else can, whatever is grand, sublime, noble, in conduct, and is thus an aid in the stimulus to the heroic.　But the safer basis for the teacher is History; Pericles, Timoleon, King Alfred, John Hampden, Grace Darling, can be depicted in the colours of sobriety and truth without detriment to their inspiring example.　The heroes of romance and poetry are most frequently impossible combinations.　A poet is either very sanguine and holds out delusive hopes, or else cynical, and distorts the legitimate expectations of the human mind. . In Romance, the personages are sure to be over-rewarded for any good they do.　A poet that would lend his genius to the vocation of moral teaching, would endeavour to be true to life, and yet

colour it with a *gentle halo* of the attractive and agree-able; and such would be the kind of composition that the instructor of youth would desiderate to assist him in his work.

We do not quarrel with our laureate for his lines in the Ode to the Duke of Wellington—

> Not once or twice in our fair island story,
> The path of duty was the way to glory—,

but we know that the glory of the Duke of Wellington demanded many other conditions than duty; and that very few, in any age, come to glory, however well they do their duty. A teacher might make a safe and sober lesson out of the Duke's career, without altogether ignoring his great personal endowments and his ad-vantages of fortune.

5. The vast theme of Mutuality in services, good offices, and affection, is inexhaustible. The purpose of self-devotion or self-sacrifice in one man is not to pander to the self-seeking of other men, but to make them enter into the relationship of mutual giving and receiving, in which human beings find their greatest happiness. One-sided devotion is temporary and pro-visional; if it does not bring the fruit of reciprocation, it naturally ceases. So great, nevertheless, is the realized blessing of genuine mutuality, that we should go great lengths to attain it.

The most salient example of the principle of mutu-ality is courtesy, or mutual kindness in small things. People can be educated thus far with comparative ease To go the length of bearing one another's heavier bur-dens is a much rarer achievement. Yet there is very little substantial virtue without this. The difficulty lies in commencing; we feel so little assured that we are not throwing away our sacrifices. The average man cannot afford to be generous when all around him are selfish.

In inculcating, as the teacher must, the duty of working for others, he should not throw overboard the reciprocity of services, as the crowning of the work. This alone keeps his pace steady, under the incitements of highly wrought ideals of self-devotion; it is the reward that is neither illusory, nor infinitely remote.

6. Humanity, in the shape of forbearance in the first place, and of active help under extreme need in the next, is the best worn topic of the lessons to the young. The tales accommodated to this end are numerous and happily told; and the iterations of it during the earliest years cannot be without fruit. Like everything else, it suffers by being ill-timed or overdone; but if a teacher is instrumental in making any moral impression at all, it should be this. True, the effect produced on tender years will be submerged in the un-tender years that follow them, but it will ultimately re-appear. It is well that the youngest should begin to feel revulsion against cruelty, oppression, intolerance; against the horrors of slavery and the brutality of despotism; and not least, against the abuse of power over the lower animals.

7. The virtue of Truth-telling calls for a special remark. The telling of a lie is an act so explicit and distinct, that it can be brought home to the offender beyond all possibility of evasion. There is, however, a defect of treatment when lying is regarded as vice standing on its own independent foundation. In point of fact, there is always a power behind that needs also to be

grappled with.   A lie is told to gain a purpose -to
evade a penalty or secure an advantage ; and we should
trace it back to its groundwork in these primary forms
of selfishness, and deal more with them than with their
formal instrument.   The admonition or the punishment
can be made much more appropriate when the insti-
gating motive is before us. A lie to escape a tyrant's rod,
is not the same as lying to snatch an unfair advantage ;
and the mode of treatment should vary with the motive.
It is only such as are fairly and kindly dealt with that
grow up truth-speaking ; in them, lies are without palliation or excuse : with others, telling the truth under all
circumstances is moral grandeur, and, when commended
as an example, should be set forth as heroic.

8. It may bring these desultory remarks to a point
to advert more particularly to some of the common mis-
carriages in well-meant moral teaching.   There are a
good many commonplaces of moral suasion that do
not bear a scrutiny ; that are either sophistical in principle, or nugatory in operation.   A few examples will
suffice.

A common lesson with children is drawn from the example of animals, especially as an incitement to industry.
The bee and the ant are supposed to put to the blush
the idle among human beings. As an agreeable exercise
of the fancy, such comparisons could be tolerated ; but
there is no suitability or relevance in likening subjects so
widely removed as human beings and insects.   There is
no record of anyone being made industrious really by
the example of the bee ; it may be reasonably doubted
whether any animal was ever adopted as a model of any
virtue, or as a beacon against any vice.   Such allusions

should never be treated as serious; they are simply fanciful and amusing ; and may easily become silly. Though
children cannot be made logical, they need not be made
illogical by false analogies. If the ant is a model for
industry, it is equally so for slave-keeping, and for all
those other questionable practices recently reported by
Sir John Lubbock as found in ant communities.

Notwithstanding that Industry is not the fulfilling of
the whole law, it is at least the basis and *sine quâ non*
of the other virtues. Hence, to reconcile the young to
abandon ease and self-indulgence for labour is one of
the most urgent topics of moral suasion. The greatness
of the stake makes it all the more incumbent on us not
to make any false move. Our incentives should be
well-grounded in fact, as well as efficacious with those
addressed. Now it seems a wrong start, to take the
very high ground, so often taken, that labour is a good
or a blessing in itself; that people cannot be happy
without it ; that the most miserable of mankind are
those that have nothing to do. A correcter view of the
necessity of labour would be equally influential as a motive. Every well-constituted human being has a certain
amount of disposable energy, to expend which gives
some pleasure, or at worst, is not painful, while health
lasts. This is to be employed in gaining a livelihood,
and as many enjoyments as possible. There is sometimes a reluctance to be overcome in employing this
amount of strength for any purpose ; and often a great
reluctance in employing it in particular ways. Yet such
reluctance has to be overcome ; nay more, the expenditure of strength must often reach the point of painful
fatigue. Still, considering that without enduring these

crosses, we cannot obtain the supply of our wants, still less the greater pleasures of life, our wisdom lies in submitting to the evil for the sake of the good. Such is the fair account of the conditions of Labour. There is great exaggeration in describing the miseries of the idle rich : we may even exaggerate the misery of the idle poor; for these rely upon others for the necessaries of life, and, in consideration of ease, can do without luxuries. This position should be assailed as grovelling, despicable, and precarious, rather than as absolutely miserable.

There is much needless and transparent sophistry in pronouncing all labour alike honourable. It is true in a certain narrow sense, that to work for one's livelihood is to enter into the common brotherhood of men, on whom, with a few exceptions, this necessity is laid. But honour means distinction, the singling out of a few among the many. For reasons, partly natural, partly conventional, some kinds of labour are highly rewarded both with money and with rank; in fact, there is an understood gradation in all services, and this cannot be smoothed down. A common soldier if he does his duty, has a certain amount of payment and esteem, but both one and the other are, and must be, very moderate indeed.

On the sad subject of Poverty, the proper line must always be to dwell upon its remediable side. In the case of those that have become old or disabled without resources, there is nothing for it but to provide support at the charge of somebody; but for those beginning life, the grand lesson to aim at is to surmount the evil. There is a certain moral courage in Burns's refrain—

'we dare be poor for a' that,' but it has little else in its
favour.   Contentment is not an unqualified virtue ; it
applies only after we have done our best.

It is in this connection that sound economical lessons
are so valuable. We have to deal, at the present day,
with vast volumes of discontent, more or less openly
displayed.   The teacher cannot avoid the topic of the
enormous inequalities of human conditions, and it will
demand some skill to keep clear of sophistry ; indeed,
if he is to follow in the commonplace tracks, he will find
sophistry plentiful.   Inequality has its first legitimate
justification in superior industry, energy, and ability ;
affluence is the fruit of an industrious career.   The inequality thus brought about is respected by everyone
but the thief.   The next stage opens up all the difficulties.   The successful man bequeaths his amassed fortune
to his children, relieving them from labour in every
shape; and the expediency of respecting property is
made to cover this case also.   Is there to be no limit
then to accumulation ?   The question is a political or
social one : the sole mode of meeting the natural discontent with prevailing inequalities, is a Social Science
discussion ; and the wants of the time require this to be
thorough.

We have viewed Morality hitherto without naming its
connection with RELIGION.   The schoolmaster of the
primary school is expected to be an instructor in Religion, both in its own proper character, and as the support
of the highest morality.   If a very formal and perfunc-
tory discharge of this duty were not readily accepted, it
would be the teacher's heaviest burden.   Any remarks

that I have to offer on so vexed a subject, will be in the same general direction as the whole of the present work, namely, the distinguishing of the diverse elements in aggregates that pass as indivisible wholes.

Some contend that Religion and Morality cannot be separated for an instant. The philosopher Kant went far to identify the two: his object was to make Morality supreme. Others think the identity necessary in order to make Religion supreme. A middle view is decidedly called for here: Morality is not Religion, and Religion is not Morality; and yet the two have points of coinci-dence. Morality cannot be the same thing without Re-ligion as with it; Religion, working in its own sphere, does not make full provision for all the moral exigencies of human life. The precepts of morality must be chiefly grounded on our human relations in this world, as known by practical experience; the motives too grow very largely out of those relations. Religion has precepts of its own, and motives of its own; and these are all the more effectively worked, when worked in separation.

¹ 'The first and most necessary instrumental for conveying ethical infor-mation to the altogether untutored, would be an ethical catechism. It ought to go before the religious catechism, and to be taught separately, and quite independent of it, and not, as is too often done, taught along with it, and thrust into it, as it were by parentheses; for it is singly on pure ethic principles that a transit can be made from virtue to religion, and when the case is otherwise, the confessions are insincere.'—KANT'S *Metaphysic of Ethics*, Semple's Translation, p. 329.

'Perhaps truth is in some degree sacrificed to system when we attempt to keep the boundary line between morality and religion clearly marked; but it holds true in general that morality is concerned with the conduct, religion with the emotions; that morality consists in the consciousness of a subjective release from all bondage. The moral precepts, which are obeyed by many, are not deduced from the religious sentiments, which are ex-perienced by few; but the connection long assumed to exist between mo

We have seen what are the difficulties and the snares of teaching, when morality is taken by itself; namely, the jumbling up of prudential, social, and heroic motives, no one class receiving its fair and full expansion. The embroilment as regards method and order must become worse by the addition of Religious doctrines and the Religious bearings of morality. It is then that we have lessons given in our schools like this—on Truthfulness: 'Truthfulness is the moral quality we transgress when we tell a lie. There is no external reward for it; it is pleasing to God; we gain a happy conscience; it is the duty we owe to our neighbour.' Such is a form of tuition prescribed by a Head Master under the London School Board.

The teaching of Religion, on its own account, is conducted in schools by means of Bible lessons, with or without a doctrinal Catechism. In many Teaching Manuals, the management of the Bible lesson is minutely formulated. In the German Schools, the order of topics in Religious instruction is officially laid down; the simple Bible narratives are given first, and the compendium of Doctrines last. If the aim of the whole were the giving of knowledge, as in ordinary instruction, the plans to follow would be such as we have been engaged in investigating all through. And, no doubt,

rality and religion is not the less real because the order of it has been inverted, for it is by the acceptance of the most abstract conclusions of morality that the mind is prepared to receive the intuitions of religion. The fruit of religious culture is a disposition to do the good without compulsion, without inducement, by an instinct that does not stop to choose or reason, and on that very account is able to override the force of impeding motives.'—*Natural Law*, by Miss EDITH SIMCOX, p. 180.

tl..re is an intellectual element in Religion ; but the essence of Religion must always be something Emotional ; and the culture of Emotion is not carried on advantageously in ordinary school teaching. The system that is best for securing the intellectual element, is not best for securing the emotional element. Regularity of lesson, method and sequence, a certain rigour of discipline—are all in favour of a steady progress in knowledge ; but the calling out and exercising of warm affection, or deep feeling, depend on improving opportunities or events such as scarcely occur in school experience. The official direction of the National Schools, in taking securities against proselytism and sectarian bias, on the part of teachers, deprives them of the freedom of action that is needed for producing emotional impressions.

In the introductory Psychological chapter, we glanced at some of the specialities attending the growth of associations with the feelings ; under the most favourable circumstances, such growths need long years to attain the intensity requisite to raise them to the pitch of high moral influence. This applies pre-eminently to the Religious feelings ; seeing that these are expected to be a power in themselves to compensate for the ills of life. We look to the wrong place, when we entrust this operation to the ordinary schoolmaster. The parent, the church, the individual's self, the spirit of the age as shown in general society and in literature,— combine to ensure the presence or the absence of the Religious tone of mind ; the national school counts for the smallest item in the total.

People might well be satisfied, as far as regards the

school, with the markedly Theistic and Christian vein of
all the Lesson-books, and with the great susceptibility
of the young mind to the explanation of the world by a
Personal God. Any results beyond, should be sought
somewhere else.

CHAPTER XIII.

ART EDUCATION.

On this subject all that I consider necessary is to define the position of Art teaching in the scheme of education —primary and secondary.

Many allusions have already been made to Fine Art. As one of the great sources of pleasure, it may be made a stimulus to study, as to any other form of effort. If education is viewed as a means of human happiness, it should not omit Art accomplishments. Moreover, among the recognized branches of common education, are Drawing, Music, Elocution, Good Breeding, Literature, —all coming within the scope of Fine Art. Once more, the environment of the school is sought to be rendered artistic.

If we are asked—Is there any Method of Art teaching ?—the answer is ' Yes.' Nevertheless, the subtleties are so great, that I shall have to be content with merely indicating what these consist in.

The Artist proper—the painter, musician, sculptor— has a mechanical and an intellectual education to go through, whose rationale is easy enough. To be a musical performer, the voice (or the hand) and the ear are trained by practice or exercise, and the general conditions of success are the same as in any other skilled exertion : natural retentiveness, the organ good, the sense delicate, iteration or practice, and, last not least, concentration induced by pleasurable emotion, or interest in the work. The *feeling* is the only speciality in Art. By the names—Taste, Æsthetic or Art sensibility, sense of Beauty—we express a complex aggregate of human emotion, which it is not easy to give account of. Art cultivation means the calling forth, intensifying, guiding, purifying, of this mass of sensibility; and it is not necessarily accompanied with the power of Art execution. The taste for music may exist without the ability to perform ; the enjoyment of pictures does not suppose the power of drawing or painting.

There can be little doubt that one way of attaining to Art-Emotion, is to become an artist. By being taught to sing, or to play on an instrument, we become versed in a wide range of musical compositions, and acquire or strengthen the taste for music. Certain original endowments are pre-supposed ; we must have the natural enjoyment of sweet sounds, a certain discrimination of tones, the feeling for concords, and perhaps some other abstruse sensibilities. These give us a pleasure in music from the beginning ; our musical education adds to, and refines upon, the primary satisfaction. An exact parallel could be given from Drawing.

But a wider view must be taken of the cultivation of the feeling for Art ; only a few are artists, the rest enjoy the works produced by these. It is considered desirable that people generally should not merely have access to performances and treasures of Art, but should be taught, or in some way assisted, to reap the full pleasure that these are fitted to afford.

To illustrate the supposed culture of æsthetic senti-

ment, I will take the example of the taste for Land·
scape. We have here a good representative of the many
different tributaries to the stream of Art enjoyment.
We must begin with certain primary sensibilities of
the senses, the eye more especially. The colour sense,
which is a very variable thing, and sometimes very
defective, must manifestly exist in a fair, if not full,
measure. The primary sensibilities to form, which are
not so easy to isolate, must also be assumed. The
emotional susceptibility to tenderness, as a chief source
of the associations of landscape, must not be wanting.
Unless I am greatly mistaken, the malevolent sentiment
also must be present, as a basis for the sublime ; although there is no need for calling the emotion into
undue exercise in its actual workings. These requisites
of sensation and emotion belong to every art ; their
natural amount, in all probability, cannot be greatly
increased ; the stress of cultivation must take some
different course.

Our next step, in the case supposed, must obviously
be to see landscapes ; and to see them in a leisurely,
deliberate manner, so as to examine their features with
care. Under this survey, the senses are gratified, the
emotional suggestiveness is brought out, and the collective interest is felt. The sight of one landscape makes
us disposed to go in search of others.

The result is greatly dependent on two circum
stances. One is a happy frame of mind ; as when we
are introduced to landscape glories, in the freshness of
youth, and in the exhilaration of holidays. This is one
of the occasions of laying up a store for the future,
by associations with pleasurable feeling. To come into

contact with scenes of beauty, in moments otherwise rendered happy, is both present and future bliss.

The second favourable circumstance is the guidance of some skilled monitor. In the presence of a beautiful scene, or a work of art, we derive great benefit from being shown where and how to direct our attention. We may chance to be misled, but it is assumed that we can find some one more advanced than ourselves in the conditions that regulate æsthetic pleasure. This is the *rôle* of the Art instructor for all.

It ought to be superfluous to remark that the landscape taste grows exactly with the devotion that we give to it. An occasional weary moment beguiled, a transient glance on the road to business, will not carry forward any taste, any rich emotional response to nature or art. We must surrender some portion of our vital energy to the accumulating of those innumerable little rills of pleasure that flow out in the presence of natural grandeur, or artistic adornment.

So far I have kept in view the main chance, how to aggrandize natural susceptibility to art pleasure, and so to create an enduring fund of delight. This is taste in the best, although not in the only, sense of the word. The cultivation of taste further implies discrimination and judgment of effects; it warns us against being pleased with certain things, on the ground that to be so pleased either interferes with the highest enjoyment of art on the whole, or brings us into collision with some of the other interests of life—as truth, utility, morality— which are liable to be sacrificed to the ends of art. This branch of æsthetics, even more than the other, needs a monitor; and taking the two together, we can see the scope that there is for Art teaching to the general community.

A short application to the chief branches of Art study, will complete the design of this chapter.

Of Music there is nothing further to be said. It is the art most universally cultivated by the mass of people; and this cultivation is followed by the taste. Without being able to perform, one may acquire the taste by listening to performances, under the favourable conditions above laid down.

Elocution is hardly yet begun to be thought of as a refining social pleasure. I shall allude to it again presently.

The group of Arts addressed to the eye—Painting, Design, Sculpture, Architecture—are the enjoyment of many; but their production is confined to a few. The culture of Taste in them has, therefore, to be carried on by the study of the works. The enjoyment increases in the manner already illustrated; while the discriminative taste may want a great deal of instruction. To reconcile all the elements in a picture, or a building, needs many adjustments and restraints that are not understood by mere natural sense, however acute.

The Poetic art raises all the questions of art culture, and on it we may expend the remainder of our observations. As being the union of language with pictures of nature and life, it exceeds the other arts in the number of elements to be reconciled. It affects a greater surface of sense and emotion than either music (by itself) or painting.

The poetic culture is involved in the course of Literature, beginning with the mother tongue. The refinements of poetry are connected with many kinds of pure composition. Every literary teacher contributes to the poetic taste, both as enjoyment and as discrimination. By reading poets and critics, under favourable auspices, we are strengthened and confirmed in the same gifts.

The susceptibility to poetry includes the ear, the eye, the emotional nature, a pretty wide experience of life, and no little book knowledge. The increasing compass of allusions in modern poetry makes it less and less the recreation of the mass of people: but there is always enough in the generality of poets to touch the chords of average human nature.

The Ideal character of Poetry was unavoidably touched in connection with moral teaching. Herein lies both its strength and its weakness. To idealize is to transcend reality or fact, and bring about a collision between ourselves and the world. The intense pleasure of the ideal is what redeems the discrepancy. The fine frenzy, the ecstasy, of the poet's world, is the inspiration to virtue, by being the spiritual reward of self-denial. It performs the part assigned to religion. Hence poets assume to themselves the vocation of being the best teachers of virtue ; Horace testifies thus for Homer, and Milton for the Greek tragedians. The question recurs—By what means do they produce their magical effects, and are these means in themselves always favourable to virtue? The poet must make himself agreeable to the multitude, and this demands concessions to human weakness ; above all things, it needs indulgences, illusions, and liberal promises. But there is much else in a true poem ; and poetic. taste and culture consists in finding pleasure apart from the exaggerations that come home to the least cultivated.

It needs little examination to discover that the strongest stimulant of Art productions is in the direction of illimitable appetite and desire—the passions of love, malevolence, ambition, sensuality. These must be stirred more or less to make the interest of a poem or a picture. The highest Art and the highest Art education check and control the outgoings of the fiery passions; hold in subjection the demons that are unavoidably raised. No greater triumph can be effected by an Artist or by an education in Art.

The two developments of poetry that have brought into greater prominence both its lights and its shadows are Fiction or Romance, and the Drama. Both have been highly popular; yet both have been inveighed against as unfavourable to morality. Although, of the two, dramatic representation is most attacked, there is but one question between them.

Fiction names a wide class; and the difference between the best and the worst examples is the whole difference between good and evil—virtue and vice. This, however, settles nothing. The crucial instances are such fictions as are most disseminated and most popular; that are imbibed by the ordinary mind without misgivings. Now, if we take the approved romances of the present day, we find much of the highest art, together with an essential tincture of indulgence in sentiment or ideality. It depends on the reader, whether the high art or the grosser element is most influential. The proper aim of Art education and culture is to enable us to feel these higher artistic effects, at the least possible expenditure of gross and grovelling passion. Such an education as this would be worthy of being promoted by all the means at our disposal.

The indulgence in the pleasures of Fiction is met by one very intelligible regulation ; namely, putting it on the level of Stimulants, to be used with moderation. The late Andrew Combe contended for a moderate use of Fiction. We can fall back upon the sober realities of life without revulsion, after a sparing allowance of Ideality, but not after excess.

The numerous works of genius that take the form of Fiction, together with Poetry in the more narrow sense, are undoubtedly an education in themselves. The force, elegance, and affluence of diction in general, the refinements and delicacies of conversational style in par-ticular, the pourtraying of character and the depicting of scenery and life, the wise maxims wittily expressed, not to mention the inspiriting ideals,—cannot go for nothing upon the mind of the reader. They are efficacious, however, just in proportion to previous culture ; with a vast majority of fiction readers, the effect is barely to be traced ; these in their haste extract only the plot, senti-ment and passion, and let all the rest escape them. To gain the full impression of a work of the highest genius demands slow perusal, and a considerable pause before entering on any other.

It seems strange that so rich a display of colloquial art as we find in the prose romance should do so little to refine the conversational part of our social intercourse. Perhaps it does more than we are aware of.

The Drama differs from Fiction in general only in

the incidents of theatrical representation.   The most obvious consequence of this is increased impressiveness; the story, sentiment or passion of the piece, whatever it is, is not changed in character, but made far more strik-ing: enhancing the goodness or the badness as the case may be.   Of course, a play is in every way more exciting than a novel, and, on Combe's rule of Temperance, should be less indulged in; but the tendency of the composition remains the same.   If our education pre-pares us to enter into the higher elements, we suffer the less from any admixture of the grosser interest.

There is one, and only one, educating influence peculiar to the Theatre, as such, and that is the art of elocution and demeanour; this too being one of the refinements of social life wherein our population is specially backward.   We see on the stage the most consummate examples of manner and address in various situations, slightly exaggerated from the necessities of distant effect, but surpassing all, except the rarest, instances in common life.   Virtue or vice may be found alike on and off the stage; but elocution and gesture can be learnt in perfection there and there alone.

CHAPTER XIV.

PROPORTIONS.

NEXT to confusion, there cannot be a greater evil than disproportion. A curriculum might be so arranged that, while each topic should be useful in itself, the whole would be a failure. It is needless to imagine absurd extremes; we shall be able to produce actual cases sufficiently glaring.

My first instance is the Cambridge system of Mathematical wranglerships. A high wrangler is a man professionally fitted for some special post involving Mathematics; but, if he turns to one of the other professions—Law, Medicine, the Church, the Public Service, he has incurred an irremediable waste of human strength.

The same remark applies to the stimulants brought to bear upon minute Classical scholarship. Unless for a professional career as a classical teacher or scholar, there is a gross disproportion here also; waiving altogether the more general question of classical study:

Natural History has not been long enough in use in the public seminaries to be abused in the same way as these two long-standing branches. But, considering that the Natural History Sciences are characterized by an interminable host of particulars, nothing is easier

than to waste time and strength over these, and to exclude the other studies that are co-essential for a broad
and liberal culture of the mind.

The error of the Cambridge wranglers may of course
be repeated in every one of the primary sciences, as
Experimental Physics, Chemistry, Physiology, Psychology. Any special encouragement given to one, co-,
operating with an individual preference, leads to the
neglect of the rest, as well as the loss of the lights that
they mutually throw upon one another. A purely psychological or metaphysical education might be the worst
case of any; in no department is it more necessary to
possess the advantages of a training in all the scientific
methods—Deduction, Induction, Classification. Logic,
which is usually coupled with Metaphysics, is not enough
of itself.

It is in Languages that disproportion, as well as
other mistakes, may be most readily exemplified. The
addition to the Mother Tongue of one foreign language,
living or dead, is a very large expenditure of mental
force, and ought not to be entered on without due calculation of probable fruits. What then shall we say to
the multiplication of languages—to the indiscriminate
imposition of two, three, or four, upon the mass of youtns
preparing for hard professional life? Very few men can
by any possibility turn to account two ancient and four
modern languages. Sir George Cornewall Lewis, the
historian Grote, and those similarly devoted to literary
and historical research, could make actual use of Greek,
Latin, French, German, and Italian; but these are rare
exceptions.

In the very limited education of young girls, it is

thought necessary to include both the French and German languages, and sometimes also Italian. No consideration is given to the likelihood of their ever taking any interest in the matter conveyed in those languages, whether as information or as polite letters. Excepting the conversational employment of them, in an occasional visit to the Continent, for which French alone is quite sufficient, the time spent on these languages seems, for the large majority, quite thrown away.

I have already alluded to the disproportional share of attention bestowed upon the purely antiquarian or archaic part of our own language. Old English is scarcely of the smallest value for the actual use of the language, and is not very generally interesting as an affair of curiosity. The proper place of Anglo-Saxon and early English seems to be among optional side-studies in the higher curriculum, and not at all in the first stages of Grammar and Composition.

There is a due proportion to be observed at every stage between information and language—between thought and expression. The excess of language chiefly attaches to the addition of too many foreign languages; we have yet to see, what no doubt we shall see, a disproportionate attention to the arts of expression in the vernacular.

In primary education, errors of disproportion are common, but not so marked and constant as in the higher curriculum. The bias of the individual teacher leads to inequalities that cannot be controlled. Moreover, sufficient attention has not yet been paid to the best selection and adjustment of topics for the needs of the pupils in after life. The primary school begins and ends the education of the masses; it begins the education of a portion of those that proceed to the secondary schools. The mixture of these two classes is at present perplexing, owing to the requirement of the dead languages in the higher education. On an amended curriculum, with knowledge and literary training as the staple, there might be an entire harmony between the primary and the secondary teaching; the course would be homogeneous throughout. The last years of the primary schooling, say from ten to thirteen, would contain systematic courses of physical and social science, with English composition and literature; to continue which would be the main task of the three or four years making up the secondary course. The sequence or progression of topics would be such that, at whatever point the pupil left school, the knowledge gained would all be available for use; there would be no wasted beginnings. Each year of the course might be made to yield the best crop that the soil will furnish. The Primary Sciences should be the goal of the knowledge course. The occupation with such tertiary products as geography and history, should not be continued longer than is requisite to prepare for the secondary strata of the Natural History subjects, and the generalized Historical or Social science; and these in turn should give place as soon as possible to Mathematics, Physics, Chemistry, &c., and, if necessary, they should be resumed in their most consummate form, by the side of the primary or mother sciences.

There can thus be no other curricular arrangement, even for the labouring population, than to give them as much methodized knowledge of the physical and the moral world, and as much literary training as their time will allow. About two-thirds of the day, as a rule, might be given to Knowledge, and one-third to Literature; music, drill, and gymnastics being counted apart from both. It seems vain to discuss any more special adaptation to the supposed exigencies of the general mass of the people.

APPENDIX.

FURTHER EXAMPLES OF THE OBJECT LESSON.

To illustrate still further the forms, conditions, and limitations of the Object Lesson, I will refer to some examples as set forth in works on Teaching. Numerous attempts have been made to give a fixed and methodical character to the lesson, for the guidance of young teachers.

Almost enough was said in the text regarding the Lessons on the Natural History type, both particular and general. The conditions in these are few and intelligible. The teacher's difficulties culminate in the third kind of Lesson—Causation, which carries us into one or other of the primary sciences. Cause and effect is at once the simplest and most impressive experience, and the most abstruse and distasteful. Nothing yields a greater charm to the child than firing off gunpowder; but, to bring anyone to the point of fully explaining the fact, needs a long course of very dry instruction. Causation cannot be excluded even from the lessons of the Natural History kind; a lesson on coal or charcoal must say something as to its combustion, but it does not (if properly managed) enter upon the theory of combustion in Chemistry. A lesson on Iron would, probably, state its melting under a high heat, but would not descant upon the laws of heat in general. One of the leading precautions regarding the Natural History Lessons is, to avoid

30

being led away to the lessons of Primary Science. It is in these last that the dangers and difficulties are of the subtlest kind, as has been partly illustrated in the text. Because we are not bound to give the full explanation of a fact—which explanation supposes a perfectly arranged sequence of topics, as in a course of Natural Philosophy, Chemistry, or Physiology— we are apt to suppose that sequence does not need to be con‧ sidered at all. We further suppose that it signifies little how many different lines of causation we enter upon in a single lesson. Again, there is a struggle between the empirical form and the rational explanation, to the detriment of both. It is only when we feel that the rational explanation is wholly beyond the powers of the pupils, that we elaborate an empirical statement with due care. The following is an example of a careful empirical statement, in dealing with the topic of Energy or Work, as measured by the raising of weights. Its object is to express the relation of velocity to height, which the pupil is supposed not to be capable of understanding on mechanical principles :—' A body shot upwards with a double velocity will mount not twice but four times as high—a body with a triple velocity not thrice but nine times as high.' Again, ' if cannon balls are directed at a compact mass of wooden planks, a ball with a double velocity will pierce four times as many planks, a ball with triple velocity nine times as many.' The more conscious we are that we cannot give the scientific reasons, the more do we aim at a precise statement of the empirical fact ; and facts so given are perhaps the very best scientific statements that can be deposited in the mind at an early age. The rule in Rhetoric, of separating, in statement, a fact from the reason of the fact, is seldom thoroughly complied with when we aim at giving both.

For further illustration, I select the following example given by an able writer on Education. The Object adduced is the Teapot Spout and Lid.

' I. THE SPOUT. Bring out that its top is higher than the

level of the teapot. What is that for? Suppose one to be lower, say half as high. What would happen when water was poured in? Give facts and illustrations so as to educe and establish the following points:—

' 1. Fluids easily yield to pressure.

' 2. Pressure is conveyed to them in all directions.

' 3. So long as there is no pressure in the spout to out-balance it the fluid must rise.

' 4. Infer that the water will rise to the same height in the spout as it is in the teapot. Illustrate also by syphon, and by pipes conveying water from reservoirs to houses.' So much for the spout. The teacher is to expand these hints as directed.

This is a lesson in the Primary Sciences, and brings out laws of Cause and Effect. As I have often urged, the teacher must settle in his mind whether it is to be an empirical or a reasoned lesson, or how far rational explanations can be carried under the circumstances; that is to say, among other things, what has been the preparation by means of previous lessons or knowledge anyhow obtained. Without being aware of these points, we cannot judge altogether of the propriety or impropriety of the lesson. In the Author's series there is no other lesson that obviously leads up to it; although in the course of explaining a great variety of familiar things, a teacher may be conscious that he has paved the way for what he now proposes to do. On the face of the scheme, we may say that too much appears to be undertaken for one lesson, and that, for so great a lesson, the start from a specified *object* is illusory. The theme of the lesson is the first chapter in the regular course of Hydrostatics, embracing the fact of fluidity under pressure, with its many consequences; and to give such a lesson effec-tively, the teacher would need to have on his table a great many objects, without distinguishing any one more than the others. The title should at once suggest the scope of the lesson : ' About Water and other Liquids and the way that they rise to their level.' In Balfour Stewart's Science Primer, we see exemplified the mode of conducting such a lesson by the help of well-chosen experiments, wherein simplification is carried much further than it could be by any teacher acting on these hints; there being, moreover, the advantage of numerous previous lessons given in a systematic course of Natural Philosophy. Many lessons on the mechanics of motion and gravity, with reference to solid bodies, should precede such a lesson on liquid bodies. It is a lesson very ill suited for isolation, even on the most rigid empirical plan. Children might be previously made to understand what is meant by 'level,' but this needs a distinct effect at explanation. The teapot and syphon lesson might be given empirically by saying that the water in the teapot, or syphon, or any similar thing, *rises to the same height* in both channels; this involves only easy notions. All explanation on first principles should be forborne, as clearly beyond the capacity of the pupils supposed to be addressed. They might be made to grasp the fact, and then be shown the consequences of pouring some more liquid into the mouth of the teapot, and at one leg of the syphon (a glass syphon with the bend downwards is the better instrument); the rise in the other leg is then seen, until the same height is reached in both. The illustration is varied and extended, when the syphon has a long leg and a short; the water poured in at the long leg flows out at the other, which no longer holds it at the proper height. The pouring out from the teapot, by lowering the spout, is then adduced as the same fact. This is a good example of an empirical object lesson on an interesting and recurring phenomenon; while at the stage when it is first given, the impression would only be spoiled by converting it into a reasoned truth, based upon the fundamental laws of motion, gravity and fluidity.

The illustration with the teapot and syphon would be enough for one lesson. A further lesson might overtake the case of pouring water at one end of a trough to see it flowing to the other end, until the surface came to rest; which might be made to appear an example of the same principle of equal height; and from this might be derived the meaning of the word 'level;' the statement being now varied by saying that water and liquids seek or find their level, or do not rest till they are level. Abstaining in the most careful manner from going back to the deductive explanation, as only competent to a class in Natural Philosophy, the teacher might expatiate upon the numerous consequences of the law in the flow of rivers, the tides, and many other facts. Probably if the mass of students of Natural Philosophy were tested, few could express strictly the deduction of the law from the fundamental laws; they have, however, been taught to express and understand it in its empirical (in strict Logic, its *derivative*) character; and only to this extent can it be communicated at an earlier stage of instruction.

I now quote from the second part of the lesson on the Teapot.

'II. HOLE IN THE LID. 1. What is a hole placed there for? Perhaps a lad will say, to let steam escape. Elicit that escape of steam would be escape of heat. What would result? Is that desirable?

' 2. Refer to boy's sucker. How is it that the stone sticks to it? Give experiments to show the pressure of the air. Infer that there is pressure of air in the spout. What would be its effect? &c.

' 3. Put two test questions. As there is a pressure through the hole in the lid, how is it that the water does not rise in the spout and overflow? The pressure in the spout balances it. When the teapot is held so that the tea runs out, what forces are acting on the spout? Two,—the pressure of the air through the hole, and the weight of the air.'

Taken by itself, this lesson is open to the same criticisms as the previous one. The preparations for it should be distinctly conceived; the limits to rational explanation ought also to be conceived, and the empirical character assumed accordingly. Further, the subject should be propounded in its real character as a lesson on the Pressure of the Air; and provision should be made for exemplifying it upon the most suitable apparatus. To connect it with a teapot is pure irrelevance. Indeed, it is something worse. For the next remark to be made is, that the two great subjects here associated with the teapot should not be broached in the same day, or even in the same month. A great many lessons on liquids ought to be given, before air is mentioned at all; and when this is entered upon, no one object is fit to be put forward as the sole or even the principal vehicle of the information. The table needs to be covered with instruments; the line of approaches needs to be carefully chalked out; a series of at least a dozen lessons must be pre-arranged. With all this, the teaching at its best can end only in an empiricism : for if the pupils are not to be trusted to educe the rise of liquids from Gravity and Liquidity, still less are they qualified to demonstrate the propositions of Aerostatics. But just as the liquid lesson culminated in an intelligible empiricism, from which many interesting natural facts could be shown to result, so it might be with the air; but not so easily. The unseen character of the agent at work makes an enormous difference. Preparing the way by enunciating the mechanical properties of the air, so as to lead up to an intelligent comprehension of the fact that it has weight, and presses upon every surface at the stated rate of 15 lbs. on the inch, what we aim at is to show what happens when the air is removed from any surface. Experimental illustrations would have to be adduced, as set forth in the Science Primers, and the empiricism distinctly stated (with no attempt to mount to the final reasoned explanation) and carefully iterated. Many lessons would be required; but in the end, the explanation of numerous interesting facts might be achieved.

PASSING EXPLANATIONS OF TERMS.

One of the most delicate parts of the teacher's work consists in explaining the meanings of the hard words occurring in
the reading lessons. The manner of doing this varies greatly.
A certain number of terms may be seen to be hopelessly unintelligible; others would occupy too much time to explain, and
are better deferred to a suitable opportunity. For those that
may be rendered intelligible, we have to consider the available
resources of explanation.

1. The Pestalozzi method of showing the Objects is the
best of all, when it can be had. This method is so obvious, but
withal of such limited application, that I do not dwell upon it.
It is not so much adapted to the school, as to the contact with
the general world, where the child is always encountering new
things, and wishing to know their names.

If a school happened to be furnished with a small museum
and a stock of apparatus, for the higher teaching, there would
be a corresponding addition to the resources for explaining
names at all stages.

2. If a thing that is already known or familiar is brought
forward by some unfamiliar term, to call the thing to mind is
to explain the term. This is exemplified in dealing with our
learned vocabulary. We are acquainted with heat and cold,
water, wind, light, under these homely terms; and when described in the more technical phraseology of the language, we
have a means of explanation in these familiar names. As instances may be cited, 'frigid zone,' a 'luminous orb,' 'auriferous
rocks,' 'aqueous vapour,' 'subterranean.'

This situation is both the facility and the snare of our word
explanations, whether in letters or in dictionaries. The defining
by synonyms is the conversion of an accident into a principle. If our language were not made up of two vocabularies,
the radical futility of the operation that underlies the method

of our dictionaries would have been discovered long ago. So completely has the idea of synonymous explanation got a hold of our minds, that we are almost as ready to give the more difficult of the equivalents to explain the easier, as the easier to explain the more difficult; ' gravity' for ' weight,' ' morose' for ' surly,' ' circumspection' for ' care,' 'rational' for ' reasonable.' It is still less surprising, that we should exchange terms that are on an equality of abstruseness: ' satisfaction' and ' gratification,' ' decorate' and ' embellish,' ' insidious' and 'treacherous,' 'proclivity' and 'inclination,' 'simultaneous' and 'contemporaneous.'

It is not enough for the teacher to guard against the error of supposing that any synonym will as a matter of course explain any other. There is a further consideration equally grave. The words called synonyms are seldom strictly equivalent; if they were so, the language ought to be disburdened of the superfluous terms. There is usually a shade of difference, and sometimes a very important distinction in meaning, between so-called synonymous terms. To interpret ' ancient' by ' old' would sometimes mislead; an ancient nation and an old nation, an ancient philosopher and an old philosopher, are very far from the same meanings. Equally wrong would be the employment of the word in rendering ' archaic ' and ' antiquated.'

3. The defect in the bare citation of a synonym is remedied by the round-about interpretation. We could explain 'ancient as belonging to the early periods of human history, and especially the period antecedent to the Christian era. A lengthened statement such as this is often necessary, and may be fully adequate to the occasion. If it contains no words or references that the hearer does not already know, and if it exactly expresses the meaning, it is a full and proper explanation. The principle implied in it is the principle of proceeding from the known to the unknown; it assumes that the thing to be defined is a combination of elements already understood. Many seemingly hard words give way under this treatment. ' Amphibious

admits of explanation to a very juvenile class; the ideas that have to be brought together, are each quite familiar. 'Temperate' is neither hot nor cold, something between the two. 'Equilibrium,' a balance, not inclining to one side or the other.

When the pupils comprehend the nature of some leading institution, as the family, or the state, they can be made to understand the various names corresponding to it in analogous institutions. Thus mother and child may be extended to the animals; the names for the sovereign power in foreign nations—Emperor, Czar, Sultan, Khan, President—are explicable when government is understood. Not only so, but the hearers can be made to comprehend important variations in the same fact or institution. Being once familiar with the Church in any one form, they can understand the other forms. An 'orchard' is explicable to those that know an ordinary garden. To 'soar' is a mode of flying.

The method necessarily fails if there be as much as one unintelligible constituent, or if there be a haziness about the elementary notions generally. A constructive operation is successful according as each one of the elements is fully grasped. 'Monopoly' cannot be understood without a considerable hold of the ideas of buying and selling. 'Revenue' demands a large preparation of political and other knowledge. 'Moralize,' occurring in a lesson in the third or fourth standards, would have to be given up as hopeless. 'Civilization' is late in being understood; its explanation belongs to a connected view of social or historical science.

Many words have successive steps of meaning, beginning with what is simple and leading on to what is abstruse. A 'mystery' may mean nothing more than is signified by concealment; it rises above this to what is in itself unknowable or incomprehensible; and finally, carries a cluster of emotions of sublimity and awe. When big words are used for their easier meanings, the task of the teacher is easy; he limits his explanation to the case. Because the word 'reason' occurs in its ordinary sense of supplying a reason or argument, we are not bound to deal with its abstruse signification in Kant's philosophy.

I doubt whether the teacher is called upon to dwell specially upon the ambiguity of words. Although many words have plurality of meanings, yet in every good composition, the ambiguity is resolved by the context, so that the difficulty is got over for the time. Continuous reading both brings out ambiguity and meets it. There should be some special reason for entering on the subject, and it should be done upon some method ; it is too wide for desultory treatment. A passing question may be allowed, after the pupils have had opportunities of encountering some word in more than one acceptation, provided always that it refers to subjects within their grasp. The word 'post' is an obvious example. So 'vice,' 'air,' 'box,' 'burn.'

4. The Figurative uses of words give a wide scope to the teacher's explanations. Here he can do much to assist the pupils, and can work in a definite line of procedure.

Figures that have by iteration lost their figurative character, and become the ordinary designations of things, as 'fortune,' 'spirit,' 'meeting,' 'court,' 'conception,' do not call for notice.

The proper Figures are those where there is an apparent stretch of meaning that to the youthful mind needs to be accounted for and justified. A '*wind* of doctrine,' a '*sea* of troubles,' 'a *surfeit* of reading,' '*stony* adversary,' '*ventilate* an opinion,' 'the *morning* of life,' 'noble *blood*,' '*simplicity* of manners,' have an effect of surprise when first heard ; and the curiosity that they awaken may be made use of to impress the meaning. This supposes that the source of the figure is something already understood. Far-fetched allusions are to be explained only under favourable circumstances.

5. It is necessary to take account of the natural or spontaneous process of ascertaining the meanings of words, after possessing a stock sufficient for understanding the drift of ordinary language. This is by a kind of tentative and induc-

live process.   On first hearing a strange word, we are often
able to judge from the connection what it is likely to mean.
Some one has committed a fault, and has been severely *cen-
sured*.   The child understands the meaning of committing the
fault, and expects some sort of penal consequence to follow ;
it is not described as punishment, yet it is something approaching it.   Perhaps it means ' scolded.'   This surmise is all that
can be made out of the occasion.   Let there, however, be
a second and a third repetition of the word :—' A writer in
the newspaper *censures* the Town Council ;' ' the Council deserves praise rather than *censure.*'   It is now plain by an induction of these additional instances, that censure is something
different from punishment; it means something painful that
we can inflict even upon superiors, and the instrumentality is
language.

We begin early this operation of guessing at the meanings
of words, from a collation of different instances, and carry it
on to the last.   What is necessary is that the general meaning
of the situation be understood.   A passage should be intelligible on the whole ; and if so, some advance will be made in
divining the sense of an outstanding word.   In rendering aid
to the struggling intelligence of the pupils, the teacher has to
meet the case thus : ' The army moved forward to engage the
enemy, and left its *baggage* in the rear with a guard.'   ' What
does baggage mean ? '   ' You see it is something belonging to
the army, not wanted immediately for fighting.'   There is no
better test of the general understanding of a passage, than the
ability to guess from it the likely meaning of an unknown word.
We could not reasonably expect the teacher to follow this up
on the instant with varied examples for inductive comparison.
Yet the operation is quite within reach, and contains the .
essence of what is meant by Induction in the highest walks of
science.

6. It must seem obvious that very important leading terms
should not be discussed under the circumstances that we are

now supposing. Such words as gravity, polarity, vibration, affinity, reciprocity, beauty, diplomacy, statute, formality, emblem, civilization,—would each form the topic of an express lesson, or else depend for their explanation upon a methodical course of their several departments of knowledge. It may so happen, however, that the purpose of their employment does not involve their most scientific use, and that they can be explained for the occasion without a rigorous definition. '*Statute* law,' or 'according to the *statute*,' could be made sufficiently intelligible for a passing allusion, although belonging to the technicalities of Jurisprudence. The word 'nature' is one of very abstruse signification, but its passing uses can often be made plain enough. The teacher, in such cases, should be aware that he is not called upon to expound such terms according to their full and exact definition. The lesson-books are somewhat misleading in this respect. The authors do not consciously make the distinction between an explanation for the purpose in hand, and a thorough, complete, and final definition. They naturally think that a word brought up in the course of a lesson should be disposed of there and then ; and that one of the purposes of the lesson is to bring forward important terms with a view to their being satisfactorily explained. This idea, pushed to the extreme, would disintegrate the lesson, and resolve the teaching into a course of dictionary work. The best and foremost use of a reading lesson is to impart a connected meaning, each part having a perceptible bearing upon every other. The portions that are clear should serve to illuminate those that are dark ; and this operation should not be interfered with by digressions for exhausting the meanings of chance terms.

There is a class of words that, occurring in this way, might be finally disposed of by one stroke of explanation, without interrupting the proper course of the lesson. They are such as are not important enough to be leading terms in science, but yet contribute to the expression of important facts or doctrines. The following are a few at random

'Salvage,' what is saved from a wreck; 'veteran,' a soldier who has been in the service long enough to have full experience, but not worn out (opposed to a fresh raw recruit); 'frontiers,' the front, or border of a country; 'retrograde,' going backwards instead of advancing, contrasted with 'progress;' 'variegated,' marked with different colours; 'reservoir,' a place where water is stored, to be run off when wanted; 'simulate,' feign or pretend to be something that we are not, with a view to deceive, while 'dissimulate' is to conceal what we are doing for the same end; opposed to both is avowing openly what we are doing.

For this class of words, the explanations in the notes to the lesson, should be careful and exact. The teacher cannot be expected to provide off hand definitions that will hit the precise points; this is the business of the lesson annotator and the dictionary maker.

7. It is useful to reflect upon the efficacy of the regular or systematic lesson in giving the correct meanings of words in entire *groups*. For example, every scientific lesson contains a number of important terms; and these occur in correlated groups. When we enter upon Geometry, we are taught tc conceive point, line, curve, triangle, square, circle, &c.—all in connection; the agreements and contrasts, and the regular sequence, make the definitions easy. Parallelogram or polygon, occurring as a passing word in a lesson, is explained, if at all, at a great disadvantage.

The same effect takes place in other instances. Thus, in a lesson on a ship, many strange terms have to be brought forward; and the quickest way of arriving at their meanings is to learn, by one continuous stroke of application, all that relates to the ship.

There are certain crafts or industries that, from being more familiar to us than others, are oftener quoted and referred to, both for information, and for figurative allusions. Such are agriculture, building, navigation, trading, criminal justice, and, perhaps above all, military operations. These have each their peculiar terms; we gather up the meanings by scattered allusions, on the tentative or inductive plan. The process might be shortened by a few compacted lessons, that would set forth in methodical array all the chief parts and processes in each department, with the appropriate designations. A lesson on the military art would be very taking to youth of ten or twelve; and would be a collateral aid to the narratives of campaigns, which are so largely drawn upon in reading manuals.

8. Although I have proposed to restrain the licence of passing explanations, by an indication of the bounds that should be set to it, thereby precluding the more elaborate and thorough modes of defining, I may still be permitted to remark, that the teacher should know what, in the last resort, thorough defining is. For all notions that are ultimate (as equality, succession, unity, duration, resistance, pain, &c.), and many that are composite or derivative, there is no definition possible except the appeal to particulars; which brings us back, after a long round, to what was said as to the mode of imparting Abstract Ideas. In lessons that are properly and strictly knowledge lessons, the handling of particulars for this great end needs to be as familiar as household words. Although its sphere is in the leading terms of science and the higher knowledge, yet it may admit of occasional application to passing terms and allusions. The word 'hallucination' could be happily explained by two or three examples, real or supposed, of persons suffering under mental delusion. So a ceremony could be illustrated by a few select instances. As these explanations necessarily occupy time, and are a new direction to the pupils' thoughts, they should be either given at the beginning of a lesson, by way of essential preparation, or be held as matters reserved, the lesson being completed by help of a mere provisional gloss.

I will now make a concluding observation as to the composition of Reading Lessons. It is desirable to exclude from these lessons, as far as possible, all terms that cause trouble to the teacher and distraction to the pupils. If a learned name occurs, intended only in one of its easier meanings, it can be dispensed with. On the other hand, it would be often useful to make room for an important term that could be readily understood in its setting, with a little assistance from the notes or the teacher.

www.ingramcontent.com/pod-product-compliance
Lightning Source LLC
Chambersburg PA
CBHW052339110726
47901CB00005B/1286